The New Wounded

forms of living

Stefanos Geroulanos and Todd Meyers, *series editors*

The New Wounded

From Neurosis to Brain Damage

Catherine Malabou

Translated by Steven Miller

FORDHAM UNIVERSITY PRESS

NEW YORK 2012

This work was originally published in French as Catherine Malabou, *Les nouveaux blesses* © Bayard Editions, 2007.

Ouvrage publié avec le concours du Ministère français chargé de la Culture–Centre National du Livre.

This work has been published with the assistance of the French Ministry of Culture–National Center for the Book.

Library of Congress Cataloging-in-Publication Data

Malabou, Catherine.
  [Nouveaux blessés. English]
  The new wounded : from neurosis to brain damage / Catherine Malabou ; translated by Steven Miller. — 1st ed.
    p. cm. — (Forms of living)
  Includes bibliographical references (p.   ) .
  ISBN 978-0-8232-3967-2 (cloth : alk. paper) — ISBN 978-0-8232-3968-9 (pbk. : alk. paper)
   1. Psychotherapy. 2. Traumatic psychoses. I. Title.
  RC480.5.M315513 2012
  616.89'14—dc23

                  2011052488

Printed in the United States of America

14   5 4 3 2

First edition

In memory of Andrée Paulhat

CONTENTS

She must think I've forgotten her since she died, how alone and abandoned she must feel! Oh! I must run and see her this very minute, I can't wait for my father to come, but where is it? How can I have forgotten the address? If only she still recognizes me! How can I have forgotten her all these months? It's dark, I won't find her, the wind is stopping me from advancing; but here is my father walking in front of me; I cry out to him: "Where's Grandmother? Tell me the address. Is she all right? Is it quite certain she's got all she needs?" "No, no," my father says to me, "you can rest assured. Her nurse is an orderly person. We send a very small sum from time to time so they can buy her the little she has need of. She sometimes asks what's become of you. She's even been told that you were going to write a book. She seemed pleased. She wiped away a tear."

— MARCEL PROUST, *Sodom and Gomorrah*[1]

There are two personal experiences at the origin of the present work. In the first place, this book is a belated reaction to the ordeal of depersonalization to which my grandmother was subjected as Alzheimer's disease operated upon her. I say "operated" because it seemed to me that my grandmother, or, at least, the new and ultimate version of her, was the work of the disease, its opus, its own sculpture. Indeed, this was not a diminished person in front of me, the same woman weaker than she used to be, lessened, spoiled. No, this was a stranger who didn't recognize me, who didn't recognize herself because she had undoubtedly never met her before. Behind the familiar halo of hair, the tone of her voice, the blue of her eyes: the absolutely incontestable presence of *someone else*. This other person, however, was strangely absent. My grandmother no longer cared about anything anymore; she was indifferent, detached, cool. In the end, she spent whole days creasing and uncreasing a corner of her blanket.

Why wasn't I comforted by this turn of events? After all, to desert life in this way, to die before being dead, isn't this the most beautiful way to die? To die to death itself? No longer to know oneself mortal? No longer to have to die *in person*? Such thoughts, however, brought me no solace. I was perfectly aware—along with everyone who must endure the same spectacle in their own lives—that this absence, this disaffection, this strangeness to oneself, were, without any possible doubt, the paradoxical signs of profound pain.

Later, I learned that Alzheimer's disease is a cerebral pathology. Could it be that the brain suffers? Could it be that this suffering manifests itself in the form of indifference to suffering? In the form of the inability to experience suffering as one's own? Could it be that there is a type of suffering that creates a new identity, the unknown identity of an unknown person who suffers? Could it be that cerebral suffering is precisely such suffering?

Another Relation to Philosophy

It took me a long time to understand how the second motive for this book, which pertains to the evolution of my relation to philosophy, is related to my grandmother's illness. For many years, my work has been devoted to the concept of "plasticity," which I encountered for the first time in Hegel's philosophy. However, the theoretical elaboration of this concept led me gradually to enlarge the field of my investigations beyond traditional philosophy into different domains of knowledge where the concept plays a decisive role—initially psychoanalysis and then cellular biology and neuroscience.

This is how I became increasingly interested in the study of the brain—its functioning, its organization, and its pathologies. This expansion of my field of research had real repercussions upon my thought, to such an extent that I can now say that there is a distinct "before" and "after" of my incursion into the domain of neuroscience. Not that I have become a "cognitivist" or a "reductionist." On one hand, I remain fundamentally attached to continental philosophy; on the other, I do not see any danger (what would be endangered by what?) in the advances of the naturalist philosophy of mind.

I consider it incontestable, from now on, that the structures and operations of the brain, far from being the glimmerless organic support of our light, are the only *reason* for processes of cognition and thought; and that there is absolutely no justification for separating mind and brain. But I was less concerned with these cognitive processes than with the affective brain, the conductors of emotion in the brain, this decisive aspect of cerebral activity that has been relegated to the shadows for too long, but that contemporary neurologists tirelessly help us to discover. There are even a large number of psychoanalysts today who know nothing or pretend to know nothing about this affective dimension of the brain. Imagine my surprise when, after giving a lecture on the relation between the psyche and the brain at a famous Parisian hospital to a room full of psychoanalysts, I was sharply chastised for misreading Lacan—who, it was objected, had already said everything I was saying.

Could it be that psychoanalysis hasn't said everything on the subject of psychic suffering? Could it be, precisely, that it ignores the suffering of the brain and, along with it, the emotive and emotional dimension of the brain?

Obviously, we can no longer consider the brain as the simple way station for stimuli without essential relation to psychic life. Alzheimer's disease, like many pathologies, is not merely a neurodegenerative disorder but also a psychic attack, in the sense that it impinges upon the identity of the subject and overturns his affective economy.

Could it be that this disease finally brings out into the open a type of lesion that psychoanalysis has never taken into account? Could it be that it manifests—a posteriori, as it were—*new forms of suffering?* Could it be that there are *new wounded* whom psychopathology has never encountered before now?

I should mention that during my grandmother's illness her geriatric care facility did not offer any psychotherapeutic services. The patients were certainly not mistreated, but it was clear that they were not considered to be subjects endowed with psychic life and that no one was prepared to respond to their despair other than by numbing it with medication. As for myself and the members of my family, we had no idea how to behave. We sat in the room, frightened, uncomprehending. We hastened to talk to my grandmother about "normal" things, as if they would still have meaning for her.

I understood too late that tenderness would have been the only way to respond; that the incoherence of my grandmother's behavior and her visible indifference were also reactions to the shock of hospitalization. If I understood more clearly, I would have tried on occasion to take her back home for a few hours. I would have given her the chance to regain her familiar surroundings, her "things." The point would not have been absurdly to help her to "refresh her memory," but to allow her calmly and without any expectations to perceive "her own absence."[2]

I did not know what to do and my books were of no help. Philosophy had even less to say than psychoanalysis. No metaphysical account of flight beyond the world had anything to do the desertion of people with brain disease. Neither the Platonic theory of the soul's aspiration to leave the body, nor the existential thinking of anxiety, the temptation of suicide, disorientation, or boredom, could shed light on this specific form of dispossession. It must be stated outright: No philosopher has ever approached the immense problem of cerebral suffering.

One must also recognize that neither psychoanalysis nor philosophy has proposed an approach to such suffering that would be at once epistemological, clinical, and metaphysical. We are supposed to be satisfied with the implicit diagnosis of vegetative state. Everyone thinks it without daring to say it aloud: my grandmother, along with all her companions in misfortune, had simply become "vegetables."

The Time of a Book

For many years, something within me quietly revolted against this type of judgment. Today, it seems that the pain caused by my grandmother's illness and the transformation of my relation to philosophy were intimately linked. My work on plasticity was perhaps a way of shedding conceptual light upon the type of psychic suffering experienced by a loved one, suffering that I could do nothing about, and that escaped the purview of the analytic categories at my disposal.

Might the new neurobiological orientation of my philosophical research on plasticity—the threefold movement of reception, donation, and annihilation of form—make it possible to recognize the importance of the cerebral psyche that is in the process of claiming its rights? To welcome, on the level of the concept, the "new wounded"? To see these sufferers as something other than figures of the unthinkable?

The unthinkable is the metamorphosis that makes an unrecognizable subject emerge from an ontologically and existentially secret place. The unthinkable is a discontinuous—most often sudden—transformation, through which a diseased identity deserts its former reference points—which it no longer recognizes as its own—and fixates upon the undecipherable touchstones of an "other world."

Might there be a type of plasticity that, under the effects of a wound, *creates a certain form of being by effacing a previously existing identity?* Might there be, in the brain, a destructive plasticity—the dark double of the positive and constructive plasticity that moulds neuronal connections? Might such plasticity make form through the annihilation of form?

I decided to begin a book in which philosophy, psychoanalysis, and contemporary neurology would enter into dialogue with one another. The tasks of such a dialogue would be to recognize and identify cerebral suffering as psychic suffering, to undertake a redefinition of the psyche itself on the basis of this recognition, and to raise the question of the brain as the source of the formations and deformations of identity.

The urgency of these tasks belongs to the future of philosophy and psychoanalysis themselves, which can no longer lag behind on questions of the psyche and the mind.

Posing the Problem

I began to write this book at the moment when the dispute between psychoanalysis and its detractors was gaining intensity. *Le livre noir de la psychanalyse* had just been published, and the media had grabbed hold of its accusations of imposture, inefficiency, fantastical definitions of the psyche, and disregard of neuronal reality.[3] The necessary responses, like that of Elisabeth Roudinesco, gave voice to legitimate indignation. But other very interesting books had also appeared that proposed to synthesize the methods of psychoanalysis and neurology, showing that the two fields were perhaps neither heterogeneous nor irreconcilable.[4] In the meantime, French philosophers remained silent about such matters.

For my part, I soon realized that nothing less would be required today than the *complete theoretical reinvention of psychopathology*; and that neither virulent attacks against psychoanalysis—as justified as they might be in certain respects—nor certain psychoanalysts' vaunted disregard for such attacks, nor the attempts at a hasty "synthesis" of the unconscious and neurons, are sufficient to accomplish such a task. Such a reinvention of psychopathology would entail both a reorientation of the clinic and the revision of the very philosophical basis of this reorientation. Before taking sides in disputes about method (analytic cure versus cognitive behavioral therapies), and before "choosing" between metaphysics and positivism, one must, humbly and rigorously, *elaborate the problem* posed by the confrontation between psychoanalysis and neurology today.

The Method: Centering and Delocalization

A problem is not a question but the elaboration of a question. This elaboration paradoxically implies both a *centering* and a *delocalization of the question* itself.

At the center of my question lies *causality*. Along the detour of the question lies *war*. At the heart of the problem—that is, at the intersection of the two directions of questioning—lies *trauma*.

Causality. It is necessary resolutely to engage with the confrontation between psychoanalysis and neurology on etiological grounds. Every psychopathology implies the elaboration of a *specific etiology*—albeit multiple and ramified—of the disturbances that it addresses. Accordingly, in order for a fruitful dialogue between psychoanalysis and neurology to be possible, one must examine the different concepts of the *causality of damage* that pertains to each science and highlight their respective understandings of the relation between *event* and *wound*.

War. From my work on the relation between neuronal architecture and social hierarchy within the capitalist enterprise, I know that any approach to psychopathology constitutes a political gesture.[5] But lengthy justifications are required to substantiate this claim. For the moment, therefore, I will limit myself to a single justification, the most pertinent to the present context. *The determination of psychic disturbances—their definition, their clinical picture, and their therapy—is always contemporaneous with a certain state or a certain age of war.*

Indeed, it is impossible to overestimate how much the present work owes to the psychiatry of war. Psychoanalysis is, above all, a theory of conflict, which was largely elaborated in proximity to the front. The role of World War I within the evolution of Freudian thought is well known. It is also well known that Freud was called as an expert witness during the Wagner-Jauregg trial in 1920. Wagner-Jauregg, a military psychiatrist, was accused of having tortured patients suffering from war neuroses under the pretext that they were "simulators," subjecting them to faradization (a type of electroshock treatment). Freud's reflections on this supposed "simulation" brought to light the unconscious signification of war and the specific anxiety that accompanies it.

My study of the relation between Freud and the logic and psychology of armed confrontation led me to read contemporary treatises on military psychology. It was undoubtedly this reading experience that, in large measure, allowed me to forge the link that I was looking to establish between psychoanalysis and neurology.

Taking into consideration changes in weaponry and the very form of military conflict in the course of the twentieth century, the contemporary psychiatry of war had been compelled, on its own, to assimilate the evolution that led from what was called *traumatic neurosis* during Freud's time to what has more recently been called PTSD, or *posttraumatic stress disorder*. War psychiatrists have a more convincing explanation for the inability of psychoanalysis to think this evolution, I believe, than *Le livre noir de la psychanalyse*. This explanation can be reduced to a single word: *trauma*. It might be—as all the conflicts of the twentieth century and the dawning twenty-first century have shown—that, for a long time now, psychoanalysis has had little of relevance to say on this subject. *Trauma thus becomes the core of the question*.

But what is the relation between war trauma and the example I began with: a patient with Alzheimer's disease? To answer this question, I have to extend my digression on war. Bruno Bettelheim's methods—by which I mean the type of gaze that he brought to the study of autistic children— are quite thought provoking. He was struck by similarities in the behavior of autistics and of "musulmans" in the concentration camps—these men who, having become indifferent to everything, let themselves die. Instead of dedicating himself to the study of autism as if it were an isolated pathology, without relation to any form of social conduct, Bettelheim asked himself whether autism was somehow a response to the threat or exercise of collective violence, a form of reaction to oppression.

Bettelheim declares: "For myself it was the German concentration camps that led me to reflect on the most personal, immediate ways on what kinds of experience can dehumanize. I had experienced being at the mercy of forces that seemed beyond one's ability to influence, and with no knowledge of whether or when the experience would end. It was an experience of living isolated from family and friends, of being severely restricted in the sending and receiving of information. At the same time I felt subject to near total manipulation by an environment that seemed focused on destroying my independent existence, if not my life."[6]

Synthesizing Bettelheim's experience with the teachings of the treatises on military psychology, it seemed that it would be legitimate to form the hypothesis that patients with Alzheimer's disease or, more generally, patients with brain lesions, behave as if they are suffering from *war trauma*.

How would it be possible not to be struck by the incontestable similarity between the behaviors of such patients and those of soldiers suffering from PTSD—Vietnam veterans (for whom the category of PTSD was devised) or, more recently, soldiers who have fought in Iraq. In particular, they all display the same affective coolness, the same desertion, the same indifference associated with a total metamorphosis of identity.

But this comparison works in both directions. Indeed, the behaviors of patients with war trauma, whether or not they suffer from patent head wounds, are comparable in every respect to those of patients with brain lesions. The work of contemporary neurologists helped me to discover *the impossibility of separating the effects of political trauma from the effects of organic trauma*. All trauma of any kind impacts the cerebral sites that conduct emotion, whether it is a matter of modifying the configuration of such sites or, more seriously, rupturing neuronal connections. Even in the absence of any patent wound, we know today that any shock, any especially strong psychological stress, or any acute anxiety, always impacts the *affective brain*— this unrecognized part of the psyche.

In order to orient a confrontation between psychoanalysis and neurology today, therefore, the first step would be the redefinition of trauma.

If there is a bridge between the cerebral and the psychic, in fact, it can only be reached by exploring the sensitive zone of the emotional brain, which constitutes a secret economy of affects and the dark core of destructive plasticity. Such an economy must be articulated with and against the traditional concept of the unconscious.

Three Hypotheses

There are three hypotheses deriving from the preceding conclusions that structure this book:

1. From sex to the brain. It is possible to deduce the existence of a *psychic regime of events—a cerebral eventality—*whose specific causality is radically different from that which psychoanalysis had elucidated. It is thus important to show—the principal wager of this book—that *cerebral eventality will replace sexual eventality within the psychopathology to come.*

2. Families of traumas. The analysis of this substitution supposes a general theory of trauma that would itself be founded upon the elucidation of *the traits that all of the new wounded have in common.*

3. Destructive plasticity. The development of the preceding points is supported by the hypothesis of destructive plasticity—until now *unknown to psychoanalysis* but also *insufficiently thematized by neurology—*that forms the psyche through the deconstitution of identity.

The goal of this book is neither to lend support to some liquidation of psychoanalysis and thus to declare unconditional devotion to the neurological approach to psychic disturbances, nor, on the contrary, to weigh down the results of neuropathology with a cumbersome theoretical apparatus. Through a sustained dialogue between the two disciplines, I simply intend to think the new faces of suffering.

Many things have changed since the period that I discuss in this book. The psychic suffering of patients with brain disorders is widely recognized today. I would like to express my debts to Dr. Thierry Gallarda (Hôpital Sainte-Anne, Paris), Dr. Laurence Lenfant (geriatric psychiatry, Dijon), and Dr. Olivier Labergère; and to thank the audiences at my lectures at Hôpital Sainte-Anne and the Cité des Sciences, the journal *L'Encéphale*, and the International Neuropsychoanalytic Society for their help and their confidence.

How could it be! An X-ray was made of my head. I, a living being, have seen my cranium—is that not something new? Come on!

— GUILLAUME APOLLINAIRE, *The New Spirit and the Poets*[7]

Introduction

The distinction between diseases of "brain" and "mind," between "neuro-logical" problems and "psychological" or "psychiatric ones," is an unfortunate cultural inheritance that permeates society and medicine. It reflects a basic ignorance of the relation between brain and mind.

— ANTONIO DAMASIO, *Descartes' Error: Emotion, Reason, and the Human Brain*

Cerebrality and Sexuality: Cause and Event

I will allow myself to invent one word and only one: *cerebrality*. My hope is that such a barbarism will come to be accepted as the mark of a concept.

Why introduce this word? It is necessary in order to construct the analogy around which my entire discussion will revolve.

FROM SEX TO SEXUALITY

Freud, as we know, distinguishes between two related ways of understand-ing "sexuality." The everyday understanding of sexuality supposes that it consists of a set of sexual practices and behaviors. The concept or scien-tific understanding of "sexuality," however, upholds it as a *law*—that is, *a specific form of causality*. Such a concept would thus function as a regulative apparatus designed to organize the phenomenal dispersion implied in the everyday understanding of sexuality.

For Freud, the ability to elucidate how this apparatus works and to establish the causal value of sexuality within the domain of mental illness—especially the neuroses—constitutes a decisive advance and will become one of the bases of psychoanalysis. To elucidate the "sexual etiology of the neuroses" is not to say that sexual problems, in the first sense, directly impinge upon the psyche—as if the latter were already constituted and incurred such lesions from the outside; it is, on the contrary, to underscore the necessary relation between such problems and the nature of psychic life itself.

Psychoanalysis does not only study "noxae that affect the sexual function itself,"[1] but also elucidates what destines or predestines these disturbances to become the styluses whereby the internal course of psychic life is inscribed. Psychoanalysis, then, is a matter of aligning the sexual etiology of the neuroses with a theory of events.

According to scientific understanding, therefore, *sexuality* appears as the concept that determines *the sense of the event within psychic life*.

FROM THE BRAIN TO CEREBRALITY

In the same way that Freud upheld the distinction between "sex" and "sexuality," it has become necessary today to postulate a distinction between "brain" and "cerebrality." If the brain designates the set of "cerebral functions," cerebrality would be the specific word for the causal value of the damage inflicted upon these functions—that is, upon their capacity to determine the course of psychic life. The recognition of cerebrality, then, implies the elucidation of the specific historicity whereby the cerebral event coincides with the psychic event. Such recognition makes possible a cerebral etiology of psychic disturbances.

If it is necessary to elaborate the concept of cerebrality today, it is because, insidiously but unmistakably, cerebrality has usurped the place of sexuality in psychopathological discourse and practice. Accordingly, this substitution is one of the basic reasons for the conflictual relation between psychoanalysis and neurology. The main purpose of my discussion will be to clarify the meaning of this substitution.

LOVE AND THE BRAIN

Although there are numerous signs that cerebrality has replaced sexuality, the relation between them has yet to be clearly articulated. All we have is the vague ideological supposition that the "brain" governs "sex." This substitution is thus expressed by recourse to the vulgar notion of an "erotic brain" at the root of all our pleasures, emotions, and suffering.

Innumerable articles—published both in specialized journals and popular magazines—posit the brain as the organ behind "sexual chemistry."[2] Such developments suggest that the hypothesis of a specific sexual drive endowed with its own psychic representation has been ousted by the idea of cerebral sensibility or sensuality. All affects begin as neuronal or hormonal processes that have different effects but derive from the same source. The libido thus gives way to the vaguer notion of "appetite," of which it would be merely one species. Accordingly, Mark Solms can declare: "Where Freud used the sexual term 'libido' to denote the mental function activated by our bodily needs of all kinds, modern neurobiologists speak of 'appetites.'"[3]

The libido—in weak sense of sexual desire or pleasure—would be merely one manifestation among others of a neuronal dynamic. The idea that the sexual would be autonomous with respect to the cerebral is thereby demoted: sex is located in the brain as one of its phenomena. As Jean-Didier Vincent puts it: "Desire is in the head."[4]

The brain has also been described as the origin of the phenomena of addiction. Advances in the knowledge of neurotransmission have made it possible to provide a detailed analysis of the processes of dependency. Whether it be drugs, alcohol, or medications—in particular, tranquilizers and antidepressants—the conclusions are the same: The habit-forming effects of these substances are due to the facilitation or inhibition of certain neurotransmitters.

According to research of this type, the brain is the origin of all of our attachments. Such an affirmation does not amount to an "intellectualization" of desire but, on the contrary, proves that cerebral organization presides over a *libidinal economy* whose laws have just begun to be explored.

Hence, a radical affirmation: Today, obscurely yet certainly, the brain appears as the *privileged site of the constitution of affects*.

EMOTIONS AND THE BRAIN

Elaborating the concept of "cerebrality" makes it possible to unify the various discourses on the brain that, even if they converge upon the same idea, remain nebulous.

What is the source of the prevalent intuition that the brain and affects are inseparable from one another? This intuition undoubtedly constitutes the naïve interpretation of what neurologists have recently called "the emotional brain." The study of emotional processes in the brain has become an increasingly important area of research within neurology that insists upon the indissoluble link between neuronal metabolism and the dynamic of emotion. Accordingly, "dynamic of emotion" does not merely refer to a system that governs a certain type of sensation but rather to a thoroughgoing redefinition of the logic of the drive.

Cerebral activity goes well beyond the mere work of cognition, and even of consciousness, to encompass the affective, sensory, and erotic fabric without which neither cognition nor consciousness would exist. For this reason, brain lesions of any kind always result in both cognitive and emotional disturbances: affective or libidinal deficits, disruption of habits, the tragic loss of life skills. Cognitive *and* emotional damage: We may now presume that the one never occurs without the other.

How, then, might we forge a consequential and rigorous connection between the "eroticization of the brain," the diffuse notion that haunts popular opinion, and the primordial importance that scientists grant to emotions within cerebral organization? The best way to respond to this question would be to examine the consequences of damage to the emotional centers of the brain. According to the available evidence, however, the type of event that interferes with cerebral affects does not fall under the jurisdiction of sexuality; it cannot be equated with "harm inflicted upon the sexual function." Brain damage constitutes a psychic event of a different nature than what Freud called a "sexual event."

The concept of cerebrality would thus make it possible to determine, much like the Freudian concept of "sexuality," both *a cause and a regime of events*. Sexuality and cerebrality appear today as concurrent economies of the psyche's exposure to wounding.

WHAT IS A PSYCHIC EVENT?

But what is the fundamental distinction between a "psychic event," understood in terms of sexual etiology, and a "psychic event," understood in terms of cerebral etiology—the distinction that establishes the differend that separates psychoanalysis and neurology? What is the difference between what happens according to the one etiology and what happens according to the other?

For Freud, as we will show at length in the second part of this book, a "psychic event" always has two sides—an "exogenous" side and an "endogenous" side. Every event implies an unexpected occurrence, an element of surprise. This is the exogenous aspect of the event. The endogenous aspect, then, comprises the way in which the psyche elaborates this exteriority in order to integrate it into the history of the subject.[5] Sexuality (in both of its senses, "empirical" and "transcendental," as it were) thus appears to Freud as the privileged site of an encounter between the exogenous and the endogenous, or, more precisely, as the privileged site of the encounter and connection between an *incident* and a *signification*. Paul Ricoeur remarks, quite correctly, that the essence of psychoanalytic discourse lies in the determination of each event as an intersection between the "energetic" and the "hermeneutic," as the connection between "nonsense" and "sense."[6]

The course and regime of events governed by cerebrality is completely different. Brain damage is itself an event that, insofar as it affects the psychic identity of the subject, reveals a certain connection between the exogenous and the endogenous. But this connection is distinguished by the fact that no interpretation of it is possible. In the case of a brain lesion, for example, the external character of the accident remains external to the psyche itself. It remains exterior to the interior. It is constitutively inassimilable.

The accidents of cerebrality are wounds that cut the thread of history, place history outside itself, suspend its course, and remain hermeneutically "irrecoverable" even though the psyche remains alive. *The cerebral accident thus reveals the ability of the subject to survive the senselessness of its own accidents.*

THE FREUDIAN REJECTION OF A CEREBRAL PSYCHE

It is precisely this psychic survival of the cerebral accident that Freud never accepted. One could even say that *his elucidation of sexuality became possible only thanks to a neutralization of cerebrality*. In his early work, Freud is indeed interested in the inscription of the event within the psyche, which, from *Project for a Scientific Psychology* onward, he elaborates as the question of "facilitation" (*Bahnung*). However, the brain will very quickly become for

Freud an opaque organ that is resistant to such questioning. An organ and an organization exposed to damage from the outside, to wounds, to lesions, to trauma, the brain remains incapable of treating the endogenous effects of such inflowing excitation. The brain is not the place where its own events are constituted. Upon this point, Freud will never waver.

This is why I decided not to rely upon Freud's neurological studies, but rather to consider his later, properly psychoanalytic texts, in which he gradually elaborates, in an ever more resolute farewell to neurology, a *new way of thinking the psychopathological event*.

The word "trauma" in Greek means "wound" and derives from *titrosko*, which means "to pierce." Trauma thus designates the wound that results from an effraction—an "effraction" that can be physical (a "patent" wound) or psychical.[7] In either case, trauma names a shock that forces open or pierces a protective barrier. However, in order to understand Freud's approach to the psychic event, one must bear in mind two fundamental postulates of the properly psychoanalytic concept of trauma: (1) The incidence of an organic lesion is incompatible with the development of a neurosis and (2) There is a distinction between the "determining" cause and the "efficient" cause of psychic trouble.

In a passage from *Beyond the Pleasure Principle* devoted to traumatic neurosis, Freud affirms: "In the case of ordinary traumatic neuroses two characteristics emerge prominently: first, that the chief weight in their causation seems to rest upon the factor of surprise, of fright; and secondly, that a wound or injury inflicted simultaneously works as a rule against the development of a neurosis."[8] When Freud wrote these lines, he was thinking primarily of cerebral lesions. Indeed, this text reprises the assertion, from "Introduction to *Psychoanalysis and the War-Neuroses*" (1919), that "the disorders that appear after frightening experiences are not neuroses."[9] Head wounds, functional and motor disturbances, paralyses, tremors, loss of memory, and disabilities that result from "tangible organic injuries to the nervous system"[10] do not, as such, have any impact upon the psyche. In order to constitute psychic events in their own right, such traumas must reactivate a conflict unrelated to war in which they happen to occur: an affective conflict. Accordingly, for Freud, the true cause of the war neuroses is, in reality, a "peacetime conflict."[11] This conflict is originally a conflict between the ego and the sexual drives: "Neuroses," Freud writes, "arise from a conflict between the ego and the sexual drives which it repudiates."[12]

The "sexual etiology of the neuroses," therefore, is only valid in cases of "traumatic neurosis" that, ultimately, cannot be truly or substantially distinguished from disturbances that result from brain lesions.

The argument is circular. Either brain lesions and other types of shock cannot be reduced to sexual disturbances, which means that patients' suffering is not considered to be psychic suffering and that they are disqualified for psychoanalytic treatment, or the aftereffects of trauma can be translated into the language of endogenous events, which means that trauma victims become common neurotics.

The distinction between "efficient cause" (*Veranlassungen*) and "determining cause"—or, more precisely, a cause that "possesses the relevant suitability to act as a determinant" (*die betreffende determinierende Eignung besitzt*)[13]—also functions to ground such an argument. This distinction, which Freud posited quite early ("The Aetiology of Hysteria," dating from 1896), allowed him to show that shocks or external events constitute only secondary factors whose damage does nothing in reality but "trigger" or "activate" the endogenous causes—the true, sexual causes—of a given disturbance.

Freud clearly underscores the need to recognize two types of factors at the origin of neurosis: constitutional factors and accidental factors. Nonetheless, when it comes to evaluating the "part played by sexuality in the etiology of the neuroses," he admits that he "ceased to lay exaggerated stress upon the accidental influencing of sexuality" (*akzidentellen Beeinflüssung der Sexualität*).[14] He also underscores the fact that if damage from outside and trauma do play a role in the etiology of the neuroses, they can only "cause secondary damage"[15] to the psyche. Although Freud was a thinker of the event, could it be that he was not a true thinker of the accident?

We must examine this point as soberly as possible. This, of course, is not to deny that Freud took external peril and threats to life into account, that he drew a distinction between fear, anxiety, and fright—the latter being the affect proper to trauma, "the state a person gets into when he has run into danger without being prepared for it."[16] He gave a perfect definition of trauma as an influx of excitation that overwhelms the metabolic capacity of the psychic apparatus. He acknowledges the status of "severe mechanical concussions . . . railway disasters or other accidents involving risk to life"[17] or "the terrible war which has just ended."[18] In the final analysis, however, sexuality, understood as a specific causality and regime of events, will always triumph over the brute accident, the pure effraction, the wound without hermeneutic future.

Did not Freud ultimately concede the point himself? *There is no beyond of the pleasure principle.*

The "development of the concept of sexuality" makes it possible to affirm the evental autonomy (*autonomie événementielle*)[19] of a process that, without entirely separating itself from the nervous system, constitutes itself as an extraterritorial site where the psychic signification of accidents integral to individual history is elaborated. Examining how Freud's path from neurology to psychoanalysis led him to change the meaning of "nervousness"— from nervous to neurotic—makes it possible to appreciate the distance that existed for him between "sexual assaults"[20] and "organic lesions" or "unexpected shocks."

For Freud, a wound such as a brain lesion is, unto itself, psychically mute. The lips of the wound must be closed to allow the "other mouth" to speak. Victims of brain lesions must be recognized as "people who are crippled in sexuality."[21]

The New Wounded

Contemporary neuropathology increasingly resists such "recognition." There are psychic accidents that cannot be translated into the language of sexual infirmity.

Cerebrality would thus designate another manner of arraigning contingency than in terms of sexuality. Although it also entails a certain interweaving of the exogenous and the endogenous, cerebrality does not allow an interpretation of the lesion or the trauma that aligns it with a "previous conflict." Neither does it accept the preeminence granted to the "internal enemy."[22] The concept of cerebrality allows for the possibility of a disastrous event that plays no role in an affective conflict supposed to precede it. Accordingly, it determines the survival of the psyche in terms of a perfectly and definitively aleatory traumatic effraction.

The sick and the traumatized people of whom I will speak in these pages have all *fallen* ill in the sense that they have been exposed to an accident, a turn of events that strikes them down in a wholly unexpected fashion. Even if they have suffered from this brutal turn as long as they can remember, if they have always been sick or will be sick forever, this does not change the fact that, in every case, the wound, permanent infirmity, or sudden lesion manifests the same *absence of sense*.

These patients, each in his or her own way, challenge us to think pure, senseless danger as an unexpected event—*incompatible with the possibility of being fantasized*. One does not fantasize a brain injury; one cannot even represent it. Cerebrality is thus the causality of a neutral and destructive accident—without reason. We know that it does not take much—a few vascular ruptures, minimal in terms of their size and scope—to alter identity, sometimes irreversibly. We know it, but the psyche cannot stage this knowledge for itself.

The impossibility of such psychic staging has psychic repercussions in itself. The psyche does indeed live, endure, and suffer from the damage caused by the "pure" accident. This damage disturbs the cerebral economy of the affects that hold together body and mind, thought and sensibility, cognition and sensuality.

The destruction of everything that attaches the subject to himself and to others—auto-affection, desire, love, hatred, pleasure—can either take an instant or the longer span of degenerative brain disease, but, in either case, the event is blind to the hermeneutic dimension. Contrary to what Freud affirms, sexuality is always exposed to a more radical regime of events: the shock and the contingency of the ruptures that sever neuronal connections.

WHO ARE THEY? PROFILES

The appellation "new wounded," therefore, designates people who suffer from psychic wounds that traditional psychoanalysis cannot understand— that is to say, understand and thus consider as relevant to its jurisdiction.

The presence of the new wounded constitutes both a return to the past and an emergent phenomenon. The "new wounded" are also the "old" wounded, people whose pathologies have long been identified. Freud the neurologist knew them on sight. At the same time, these pathologies are "new" to the extent that we are beginning today to appreciate not only their organic but also their psychic effects. Herein lies the new phenomenon: From now on, *people with brain lesions will form an integral part of the psychopathological landscape.*

Who are they? They are, as the term indicates, victims of various cerebral lesions or attacks, head trauma, tumors, encephalitis, or meningoencephalitis. Patients with degenerative brain diseases such as Parkinson's or Alzheimer's also fall into this category. In addition, we might think of the patients whom psychoanalysis has attempted to cure without success: schizophrenics, autistics, epileptics, victims of Tourette's syndrome.

The "new wounded" constitute an emergent phenomenon, then, to the extent that this category also refers to subjects who suffer from disturbances that had yet to be identified during Freud's time. For example, one might adduce several recently discovered disorders: obsessive-compulsive disturbances, hyperactivity syndrome with attention deficit disorder, or any of the illnesses identified by the "disabilities movement."

All such people—victims of accidental lesions or chronic illness—suffer, no matter their disparate clinical profiles, from emotional disturbances that essentially consist in the malfunctioning of affective signals necessary to make decisions. To differing degrees, they all display permanent or temporary behaviors of *indifference* or *disaffection*.

TOWARD A GENERAL THEORY OF TRAUMA

The "new wounded," however, are not merely people with brain lesions. We should recall that cerebrality designates a regime of eventality that recognizes the psychical weight of accidents stripped of any signification. I thus authorize myself also to extend the category of "new wounded" to cover every patient in a *state of shock* who, without having suffered brain lesions, has seen his or her neuronal organization and psychic equilibrium permanently changed by trauma. Such patients also suffer, in particular, from an emotional deficit.

The approach of contemporary neuropathology makes it possible to elaborate a legitimate model for understanding the structure of every type of psychic trauma. The "lesion method,"[23] to borrow one of Damasio's formulations, is thus capable of showing that both subjects with brain lesions and those who have suffered types of trauma not linked to cerebral pathology present identical behaviors. The behavior of subjects who are victims of trauma linked to mistreatment, war, terrorist attacks, captivity, or sexual abuse display striking resemblances with subjects who have suffered brain damage. It is possible to name these traumas "sociopolitical traumas."

Under this generic term, one should group all damage caused by extreme relational violence. Today, however, the border that separates organic trauma and sociopolitical trauma is increasingly porous.

This affirmation tends to generalize and enlarge the concept of brain damage opening it to types of harm that do not initially pertain to neuropathology. It is thus necessary to show that all trauma impacts neuronal organization, particularly the sites of emotional inductors. This is precisely the point that makes it possible to construct a paradigm for all the "new" wounded. In addition, this affirmation makes it possible to understand neuronal disturbance in other terms than pure and simple physiological lesions.

Of course, in neuropathological cases, neuronal changes are the cause of psychic disorganization, whereas they are the consequence of psychic disorganization in cases of sociopolitical trauma. Nonetheless, in all of these situations, the same *impact of the event* is at work, the same *economy of the accident*, the same relation between the psyche and catastrophe.

Not all traumas, lesional or sociopolitical, are always fortuitous, and there is never a simple relation between the "normal" interior of the psyche and the violent irruption of an unpredictable exterior. Sociopolitical trauma never occurs entirely by chance. Every event always derives, in one way or another, from an indivisible intimacy between the outside and the inside. Nonetheless, today, traumatic events appear more and more clearly as events that tend to mask their intentionality, taking two, apparently contradictory, forms: they appear either as perfectly unmotivated accidents or as the necessary blindness of natural laws. In both cases, the intentional orientation of the event is disguised or absent.

The victims of neuropathological trauma thus display a strange phenomenon that constitutes a structural trait of all posttraumatic behavior. Effacing the limits that separate "neurobiology" from "sociopathy," brain damage tends also to blur the boundaries between history and nature; and, at the same time, it reveals the fact that political oppression, today, itself assumes the guise of a traumatic blow stripped of all justification.

NEUROPSYCHOANALYSIS

Beyond the controversies and discussions about the efficacy or scientificity of psychoanalysis that currently divide the field of psychopathology, we should, instead, devote our attention to *the change in the concepts of event, wound, and trauma.*

This gesture does not entail taking sides against Freud. On the contrary, I will undertake a reading of Freud that will turn upon the elaboration of the notion of the psychic event. In order to facilitate a confrontation between Freudian thinking and the contemporary neurological thinking of the event, I will follow the path opened by the disciplinary formation of neuropsychoanalysis.

Born in the United States from the work of neurologist and psychoanalyst Mark Solms, this discipline presents itself as a novel synthesis of the neurological and psychoanalytic approaches to mental disturbances.[24] A relatively recent development, having only emerged at the beginning of the 1990s, neuropsychoanalysis, as its name indicates, is a bridge concept, a hyphen, between neurology and psychoanalysis.[25]

The neuropsychoanalysts belong to the new generation of researchers who have contested the pure and simple rejection of psychoanalysis, which, since the 1950s, has been common currency among theoreticians of the brain.[26] The scientists who work under the banner of neuropsychoanalysis uphold the necessity of returning to Freud without thereby eschewing the principle of the cerebral etiology of mental disturbances. In their view, Freud is no longer a traitor to neurology but, on the contrary, the figure who laid the groundwork for the completion of its work by outlining within the field of psychoanalysis a series of investigations and results that would be confirmed by the neurology of the future. This is the theory of the "moment of transition."

In his book, *The Brain and the Inner World*, Solms cites the famous passage from *Beyond the Pleasure Principle* in which Freud declares: "Biology is truly the land of unlimited possibilities. We may expect it to give us the most surprising information and we cannot guess what answers it will return in a few dozen years. . . . They may be of a kind which will blow away the whole of our artificial structure of hypotheses."[27] Solms refers to this passage in order to affirm that "it is not a matter of proving that Freud was right, but of finishing his work,"[28] thereby claiming that psychoanalysis took no more than a provisional distance from neurology as it waited for neurologists accept and respond to the hypothesis of the unconscious.

According to the theory of the "moment of transition," psychoanalysis would only be an episode—a sort of necessary interruption—in the long history of neurology, an episode that would mediate between the classical neurology that Freud the medical student would have learned and practiced it and the present state of the psychobiological science of neurons. "The reason [that Freud abandoned neurology]," Oliver Sacks writes, "was the very inadequate state of neurological (and physiological) understanding at the time, not any turning against neurological explanation in principle."[29] Further: "Freud knew that any attempt to bring together psychoanalysis and neurology would be premature (although he made a last attempt at this in his 1895 'Project,' which he left unpublished in his lifetime)."[30]

If this "moment of transition" is now complete, the project of conjoining psychoanalysis and neurology has yet to be accomplished. To integrate psychoanalysis within the history of neurology is at once to insist on the need to renew, or even reform, certain aspects of psychoanalytic theory in light of neurological findings and, inversely, to recognize what a neurology that has entered the "era of subtlety"[31] owes to Freud.

The concept of neuropsychoanalysis, as Sacks explains once again, initially derived from the encounter between psychoanalysis and neuropsychology—hence the name "depth neuropsychology" that Mark Solms sometimes uses for the discipline.[32] It was the great Soviet psychologist Alexander Luria, who, advancing the work of his master, Lev Vygotsky, founded neuropsychology in the 1930s. Luria proposed to replace the notion of "cerebral function" with the concept of "functional system."[33] While the function is anatomically located in an "air," the "functional systems" suppose dynamic interactions between different neuronal mechanisms. These systems are characterized, in particular, by the ability to reorganize their elements; and this means that a lesion does not merely affect a single place in the neuronal organization but transforms the linkages or interactions between the systems. Brain lesions always have a dynamic localization.

It was in this way, Oliver Sacks declares, that

> neurology itself had to evolve, from a mechanical science that thought in terms
> of fixed "functions" and "centers," a sort of successor to phrenology, through
> much more sophisticated clinical approaches and deeper understandings, to
> a more dynamic analysis of neurological difficulties in terms of functional
> systems, often distributed widely through the brain and in continual interaction
> with each other. Such an approach was pioneered by A. R. Luria in the Soviet
> Union. But neuropsychology, as this approach came to be called, only got going

during the Second World War, so, sadly, Freud never saw it, never knew how Luria had lifted clinical psychology to an entirely new level, a level perhaps complementary to that of psychoanalysis.[34]

Conceiving cerebral organization in terms of functional systems makes it possible to understand how brain lesions can affect several functions at the same time and interfere with the patient's entire identity.

In his introduction to *The Man Who Mistook His Wife for a Hat*, however, Sacks explains that Luria primarily took into account the repercussions of lesional damage upon systems of cognitive function, or the "higher cortical functions of man."[35] Neuropsychoanalysis, however, assumes the task of studying the repercussions of lesional damage upon the brain's systems of emotional and affective functioning. Accordingly, the approach of neuropsychoanalytic practice is both "to make the most detailed neuropsychological examination of patients with brain lesions and then to submit them to a model psychoanalysis."[36]

It is thus necessary from a neuropsychoanalytic viewpoint "to bring the mechanisms of the brain and the inner world of the patient together."[37] Neuropsychoanalysis recognizes the essential and reciprocal link between the life of the brain and subjective experience.

Why, then, didn't I subtitle this book "Freud and Neuropsychoanalysis"? It is important to underscore the fact that neuropsychoanalysis is not a constituted science but a working discipline, an experimental inflexion within contemporary neurology.[38] It should not, therefore, take center stage. It is more important to utilize the aspects of neuropsychoanalysis that help to stage the confrontation that I venture in these pages between two regimes of the event: sexuality and cerebrality. The fundamental lesson of neuropsychoanalysis consists, from now on, in the imperative to place "sexual cripples" and "survivors of neurological disease" on the same level.[39]

Plasticity and Changes in Personality

The concept of the event—understood as a traumatic event—allows for a decisive critical distinction between psychoanalysis and neurology. Every wound, whether "sexual" or "cerebral"—if one agrees to understand these words on the basis of the etiological value conferred upon them—has the power to change the personality of the patient. But psychoanalysis and neurology each endow this change with a profoundly different sense.

PSYCHOANALYSIS AND NEUROLOGY DO NOT ENTAIL
THE SAME IDEA OF CHANGE

What does this mean? To describe phenomena linked to the transformation of identity, Freud uses the word *Ichveränderungen*—"alterations of the ego" or "modifications of the ego."[40] Profound as they may be, such "alterations" or "modifications" never entail an absolute rupture with the patient's foregoing personality. Changes caused by brain lesions, however, frequently manifest themselves as an *unprecedented metamorphosis* of the patient's identity. "Unprecedented," in such cases, signifies "without any relation to the subject's past": the wound gives rise to a *new person*, precisely, to one of the new wounded. A person with Alzheimer's disease, for example, is not—or not only—someone who has "changed" or been "modified," but rather *a subject who has become someone else.*

All lesions that impact the cerebral mechanisms for producing and regulating emotions (particularly in the prefrontal cortex, the hippocampus and the amygdala) can alter the personality to such a degree that it becomes unrecognizable without necessarily diminishing the higher cognitive functions (language, memory, attention, and so on). This alteration manifests itself especially in the "odd unconcern"[41] that seems to come over the new wounded, as if they had been separated from themselves.

This "change in personality" thus designates such a disruption of identity that it, or the wound that causes it, constitutes a bright dividing line between "before" and "after." Such radical change corresponds to the definition of catastrophe proposed by the psychiatrist and psychoanalyst François Lebigot, a specialist in emergency situations: "Catastrophe defines the event as it asserts itself on the psychic level; that is, it represents a localizable and most often brutal external fact that, for the subject, causes a rupture which introduces a very radical division between before and after."[42]

PHINEAS GAGE

The paradigmatic example of such change, cited in many neurological studies, is that of Phineas Gage.[43] A railroad construction foreman in Vermont at the end of the nineteenth century, Phineas Gage was directing a rock-blasting operation when the accident happened. He triggered an explosion as he was compacting a charge in a rock formation with a long iron rod. The force of the blast drove the rod all the way through his skull.

Miraculously, he survived the accident, but his frontal lobe was gravely damaged. Gage became both irritable and indifferent to everything. Having lost any feelings for his friends and family, he seemed utterly disaffected. Damasio writes, "Phineas Gage will be pronounced cured in less than two months. Yet this astonishing outcome pales in comparison with the extraordinary turn that Gage's personality is about to undergo. Gage's disposition, his likes and dislikes, his dreams and aspirations are all to change. Gage's body may be alive and well, but there is a new spirit animating it."[44] And later: "Gage was no longer Gage."[45]

Because this metamorphosis was a metamorphosis of being in its entirety rather than a mere loss of aptitude, it is not possible to separate the organic wound from its psychic repercussions. Mark Solms, who claims to have treated "hundreds of Phineas Gages," affirms: "Today we know, from observing countless similar cases, that damage to that area of tissue almost always produces the very same type of personality change that it did in Gage's case."[46] Indeed, "there is a predictable relation between specific brain events and specific aspects of *who we are*. If any one of us were to suffer the same lesion in that specific area, we would be changed in much the same way that Gage was, and we, too, would no longer be our former selves. This is the basis of our view that anyone with a serious interest in the inner life of the mind should also be interested in the brain, and vice versa."[47]

THE "AGONIZING QUESTION OF DISAFFECTION"
AND THE POSTTRAUMATIC CONDITION

The reason for this "interest" resides in the psychic capacity for metamorphosis at work in cases of brain damage. What is the origin of this capacity? Why does it raise the "agonizing question of disaffection" that accompanies trauma?[48] Why is disaffection the most common posttraumatic symptom? Françoise Davoine and Jean-Max Gaudillière, psychoanalysts who are open to neurological problems, affirm that their "clinical work brings us into contact with many of Phineas Gage's descendants, as it were—people whose cortex was not necessarily mutilated but whose emotions were nevertheless disrupted and anesthetized, leading to horrific decisions."⁴⁹ Behavioral disaffection responds to the cool indifference of the forces that cause trauma. Destructive metamorphosis is the effect that responds to the implacable and senseless character of the cause.

The new wounded, people with brain lesions, have replaced the possessed or the madmen of ancient medicine and the neurotics of psychoanalysis. The specter of such phenomena hints at the scope of a *posttraumatic condition* that reigns everywhere today and demands to be thought.

PLASTICITY AND DESTRUCTION

These remarks bring me to my third main idea: *the apparition of a new face of plasticity.*

To recognize the determining causal value of the wound is to take into account its *plastic* power upon the psyche. The term "plasticity," one should recall, has three principal significations. On one hand, it designates the capacity of certain materials, such as clay or plaster, to receive form. On the other hand, it designates the power to give form—the power of a sculptor or a plastic surgeon. But, finally, it also refers to the possibility of the deflagration or explosion of every form—as when one speaks of "plastique," "plastic explosive," or, in French, *plastiquage* (which simply means "bombing"). The notion of plasticity is thus situated at both extremes of the creation and destruction of form.⁵⁰

Which of these three senses should one retain to characterize the plastic power of the wound upon the psyche? Certainly, this power is the power to create form, in the sense that it brings about a metamorphosis of identity. And this identity is itself plastic to the extent that it is susceptible to being imprinted by this new form. Nonetheless, it is clear that wounds—traumas or catastrophes—are not "creators of form" in the positive sense of the term. We are quite far from the sculptural paradigm of "beautiful form." If the wound, as the determining cause of the transformation of the psyche, has a plastic power, it can only be understood in terms of the third sense of plasticity: explosion and annihilation. If brain damage creates a new identity, this creation can be only *creation through the destruction of form.* The plasticity at stake here is thus destructive plasticity.

Such plasticity—and herein resides its paradox—ultimately remains an adventure of form. What patients with Alzheimer's disease show us, to take this example once again, is precisely the plasticity of the wound through which the permanent dislocation of one identity forms another identity— an identity that is neither the sublation nor the compensatory replica of the old form, but rather, literally, a form of destruction. Such patients prove that destruction is a form that forms, that destruction might indeed constitute a form of psychic life. The formative-destructive power of the wound, as we are attempting to think it now, may thus be articulated in this way: *All suffering is formative of the identity that endures it.*

How, then, does the theme of destructive plasticity orient the critical confrontation between psychoanalysis and neurology that I am now undertaking? To begin with, it must be acknowledged that neither Freud nor the neurologists have elaborated the concept. In psychoanalysis and in neurology, plasticity is a powerful operative category, but it is only ever understood in terms of its first two senses: reception and donation of form. The third sense—that of deflagration—is ignored. Neurological reflection upon the determining power of the wound and trauma is certainly a reflection upon the change in identity that destroys this same identity. But this reflection lacks a *concept* that would make it possible to define the meaning of this change and to grasp the psyche in terms of its capacity to survive after the wound—not as absence of form but as the form of its absence. If the category of plasticity does play a role in both psychoanalysis and neurology, it gives no more than a hint of its own negativity.

On Freudian Plasticity. What does this mean? In Freud, "plasticity" mainly designates two essential phenomena: on one hand, *the vitality of the libido*, its capacity to change its object and to resist fixation; and on the other hand, *the indestructible character of psychic life*. Within the psyche, nothing is forgotten; traces have an indestructible character. Imprints can be modified, deformed, and reformed—but they persist. A very beautiful passage from "Thoughts for the Times on War and Death" elaborates this sense of plasticity:

> The development of the mind shows a peculiarity which is present in no other
> developmental process. . . . Here one can describe the state of affairs, which has
> nothing to compare with it, only by saying that in this case every earlier stage
> of development persists alongside the later stage which has arisen from it; here
> succession also involves co-existence, although it is to the same materials that
> the whole series of transformations has applied. The earlier mental state may

not have manifested itself for years, but nonetheless it is so far present that it may at any time again become the mode of expression of the forces in the mind, and indeed the only one, as though all later developments had been annulled or undone. This extraordinary plasticity (*dieser ausserordentliche Plastizität*) of mental developments is not unrestricted as regards direction. . . . But the primitive stages can always be re-established; the primitive mind is, in the fullest meaning of the word, imperishable.[51]

This "extraordinary plasticity of psychic developments" thus designates the character of what resists destruction and forgetting, albeit at the price of regression.[52] Plasticity displaces without annihilating. Indestructibility is the rule in psychic life and the norm of psychopathology. "What are called mental diseases," Freud writes, "inevitably produce an impression in the layman that intellectual and mental life have been exposed to destruction (*Zerstörung*). In reality, the destruction only applies to later acquisitions and developments. The essence of mental disease lies in a return to earlier states of affective life and of functioning."[53]

However, the study of the contemporary neurology of brain wounds and traumas raises fundamental questions: Can we be sure that psychic life resists destruction? Can we be sure that there is something indestructible about the psyche? Can we be sure that the set of "endogenous events" that constitute sexuality, in the scientific sense, resist attacks coming from outside, do not succumb to explosion or annihilation? Can we be sure, finally, that we can still consider "mental illness" as a "return to earlier states"?

The hypothesis of destructive psychic plasticity, therefore, calls into question the idea of the continuity of personality in pathology. It is entirely possible that there will be *no relation* between the identity that comes before a lesion or trauma and the identity that comes after, that, once again, the new identity will be unprecedented. Coolness, to come back to it, is certainly the most convincing argument in favor of such a metamorphosis. Damasio writes about one of his patients: "[His] emotional life seemed impoverished. Now and then he might have a short-lived burst of emotion, but for the most part such display was lacking. There is no sign that he felt for others, and no sign of embarrassment, sadness, or anguish at such a tragic turn of events. His overall affect is best captured as 'shallow.'"[54]

On Neurological Plasticity. Nonetheless, when neurologists speak of a person becoming unrecognizable, even if they invoke the archetypical figure of the cold-blooded killer, they do not truly theorize this negative plasticity. Between the traumatic effraction and the response of identity to this effraction, there remains a space of *psychic elaboration*—a space that is never explored as such even though it constitutes the veritable site for a fruitful engagement with psychoanalysis.

The neurological concept of plasticity itself also remains attached to the positive values of neuronal construction and configuration, of the creation of a style of being. The two types of cerebral plasticity, constructive and destructive, are never related to each other. Both cases, however, entail an elaboration of form. How can these two plasticities coexist?

All of these questions are questions that I address in this book. It has often been objected to me, in spite of my reiterated insistence upon the three senses of plasticity—reception, donation, and annihilation of form— that, ultimately, I myself privilege the first two (creative) senses over the final (negative) sense; that I merely evoke destructive plasticity without ever concretely envisaging it; that I only ever deal with it allusively; that I only ever explore the creative dimensions of plasticity: invention, suppleness, resistance, the ability to oppose flexibility. It is true that in the final analysis, I have never made the senselessness of form or the annihilation of sense in form into the thematic focus of my work. Never have I truly confronted the possibility of destructive metamorphosis.

This is the book in which I would like to go "a step farther," to explore the hypothesis of a truly explosive plasticity and to stop inscribing deformation within the project of sublating form.

Thus it is in this direction, as unknown to me as to psychoanalysis and neurology, that I would like to open a dialogue, developing the idea of a plasticity that would be nothing other than a form of death.

The plastics of death: How should one think this idea? This is perhaps the most difficult problem, the most vulnerable in my work, because it must be situated between the Freudian hypothesis of the death drive—which infinitely complicates the problem of sexuality—and the contemporary neurological hypothesis of a death of the drive.

PART ONE

The Neurological Subordination of Sexuality

Introduction: The "New Maps" of Causality

The field of psychotherapy has always been a site of ideological confrontation.

— DANIEL WIDLÖCHER, *Les nouvelles cartes de la psychanalyse*

Two Causalities: System and Alea

In the *Critique of Pure Reason,* Kant shows that there can be no causality without a "character." And "character" is defined as the "law of a causality without which it would not be a cause at all"; that is, as the definite and specific relation between a cause and its effects.[1] A cause cannot be a cause unless it presides over a precise order of events. Accordingly, if sexuality and cerebrality are each "efficient causes," they each must "have a character" and intervene within a specific field as instances that regulate events, that have the capacity to condition relations between phenomena.

The confrontation that I would like to establish between these two etiological characters, therefore, seeks to cut short any fruitless attempt to establish "bridges" between "neuronal states" and "psychological states." To borrow a clarifying formulation from Daniel Widlöcher, such a connection between phenomena that pertain to different causal regimes always lead to the aporia of an "impossible parallelism." Rather than extract snapshots from the flux of causal series in order to compare them, I will put these series themselves into perspective. I will be concerned with "events" rather than "states."

The Struggle for Etiological Dominance

The event, it would seem, raises the same question for both the psychoanalyst and for the neurologist. Does the disturbance or lesion derive from an internal disorganization of a system, from an endogenous disorder? Or must we admit on the contrary that this disorganization has been introduced by external events? It is in the nature of any causal "character" to synthesize the *systematicity of the system* and the *happening of the accident*.

As Marc Jeannerod quite rightly asserts, the psychoanalyst and the neurologist must both construct functional systems.[2] The psychical apparatus, in Freud, is composed of instances forming an internal organization that determines the meaning of the event. Neuronal architecture is itself composed of different systems that are constantly interacting with one another.

Nonetheless, it is just as impossible for the psychoanalyst as for the neurologist to adhere to a strictly functionalist explanation of psychic disturbances that would take into account only the systematic level, ignoring external events and thus "reducing exogenous causes to nothing but factors that reveal vulnerability."[3] According to Jeannerod, the necessary collaboration between psychiatry and neuroscience today must be based upon their shared understanding of the way in which exogenous causes impact intrapsychic mechanisms.[4]

In *La causalité psychique*, André Green also distinguishes between a "closed system" and an "open system": a system of functions is at once an "auto-organization" or an "autopoesis" and a structure open to an "event."[5] This is perhaps a translation of what Freud formulates in *Leonardo da Vinci and a Memory from His Childhood* as one of the fundamental principles of psychoanalysis: "Our aim remains that of demonstrating the connection along the path of libidinal activity (*über den Weg der Triebtätigung*) between external events (*äusseren Erlebnissen*) and a person's reactions (*Reaktionen der Person*)."[6]

Psychoanalysis and neurology thus share the task of thinking the psyche as something that entails both the autoregulation of the system and the intrusion of the alea, both economic necessity and the irreducible margin of indetermination.

This shared task, however, immediately appears as a point of rupture. The psychoanalytic and neurological conceptions of the relation between system and accident, along with their conceptions of the event itself, are radically opposed to one another. It would even be possible to claim that the conflicts between these sciences result precisely from their incompatible approaches to psychic eventality. This incompatibility might explain, in large part, the long process of mutual neutralization that has structured both psychoanalysis and neurology since their respective beginnings. Sexuality and cerebrality have always waged a fratricidal war for etiological dominance.

Today, the etiological value of sexuality has been subordinated to that of cerebrality, thereby reversing the old hierarchy that, until quite recently, privileged psychoanalysis.

"Subordination of sexuality" should be understood to mean, first, the effacement of the specificity of the sexual within the affects governed by the emotional activity of the brain. Second, it should be understood to refer to the ever-greater etiological predominance of cerebrality within psychopathology. These phenomena are contemporary with a profound redefinition of the psychic event.

The psychic event will no longer be considered as a sexual event. The hypothesis of the emotional brain ousts the idea of an autonomous sexual drive. The regulation of events by the *cerebral character* thus differs profoundly from the regulation of events by the *sexual character*.

Calling "Psychic Energy" into Question

Let us clearly articulate the dividing line: Contemporary neurology fundamentally contests the concept and the existence of "psychic energy," which constitutes, for Freud, the motor that regulates all the incidents that impinge upon the psyche—that is, everything that happens.

What does this mean? Eventality, whether cerebral or sexual, always supposes an energetics. Neurologists, however, reject the hypothesis of a "psychic" energy that could be separated from nervous energy. The neurological theory of the emotions thus makes it possible to conceive affect without distinguishing between psychic and neuronal energetics.

For Freud, as we will see, *the brain is not in a position to treat—that is, to regulate—its own endogenous excitation.* For this reason, it is also not in a position to bind excessive excitation coming from outside. Psychic energy thus takes up where neuronal energy leaves off in order to divert internal and external stimuli and to regulate the encounter with them within a specifically constituted space—the *psychic apparatus* that is the sole instance in a position, at the site of this encounter, to constitute the event.

The libido—as a "dynamic manifestation" of the "sexual drive"—is the most constant manifestation of psychic energy. It is characterized by its flexibility and great powers of displacement. The "libido" thus allows Freud to define psychic energy in general. If psychic energy indeed results from the displacement of nervous energy, from its channeling into a extracerebral site, then any psychic event whatsoever is always libidinal to the extent that it has been constituted through such displacement. The sexual is the detour of the cerebral.

The neurological subordination of sexuality results from calling into question this energetic organization within which the libido, thanks to its fluidity, supplements the brain's inability to support its own excitation. There is only one type of energy: nervous energy. There is only one type of energetic circuit: neuronal dynamics. The often-implicit thesis that structures the theory of the emotional brain could thus be formulated in this manner: the brain is an auto-affectable instance that is perfectly capable of managing both internal and external disruptions. The cerebral economy of emotions does not require the diversion of psychic energy and the constitution of an apparatus whose imaginary topology would superimpose itself, in some sense, upon the anatomical and biological organization of the nervous system. Accordingly, the neurological subordination of sexuality goes hand in hand with the postulate of cerebral autonomy in the treatment of affects, an autonomy that necessarily calls into question the pertinence of the concepts of drive and libido. *The brain has its own events.*

In her article "Psychoanalysis at the Risk of Neuroscience," published in *Le livre noir de la psychanalyse,* Joëlle Proust contests precisely the Freudian concept of an excitation other than nervous excitation that would be the true source of all psychic energy, *including nervous energy.* She declares:

"The Freudian theory of the relation between the psychic and the somatic supposes a certain conception according to which the neurons receive their excitation from outside. Freud concludes that there must be a 'peripheral' somatic excitation in order for the nervous apparatus to be stimulated. Nervous influx is understood to be a form of energy that streams through the neurons but is not engendered by them. This energy then invests itself—that is, attaches itself to certain representations that become the representations of the corresponding drive."[7] This energy coming from outside is nothing other than the libido, "which can be excessive, insufficient, dammed up in one place, free-flowing in another. According to Freud, the different vicissitudes of the drive explain the formation of complexes such as the Oedipus complex and their pathogenic developments, such as neurosis and psychosis. However, we know today that mental energy does not have an extraneuronal origin: the axon of the neuron produces the nervous influx that streams toward the nerve endings. The idea that the libido organizes psychic life is thus stripped of its only neurological justification. This discovery makes it necessary to revise the image—and the theory—of the reservoir of energy that Freud thought to be provided by the somatic drives. By the same token, the drive theory of mental illness—the role supposed to be played by the complexes and the return of repressed representations—would also be deprived of its former significance."[8]

This argument, which brutally synthesizes many quite distinct levels of Freudian analysis and only provides the crudest account of drive theory, still deserves credit for seeking a new approach to psychic energy and neuronal energy; to the internal and the external; to the nervous system and its relation to the outside, the accident, or the event. The major line of argumentation, which does away with the libido and the drive, hinges upon the affirmation that the brain itself regulates what happens to it without outside help or intervention. If this is the case, then it can no longer be supposed that sexuality occupies a terrain apart from the cerebral.

The very particular manner in which the brain affects itself—noiselessly, without tangible exterior signs, only occurring by accident in the absolute fragility of its unconscious exposure—will be at the center the following analysis, which anticipates its own a posteriori Freudian critique.

ONE

Cerebral Auto-Affection

What Freud quite awkwardly calls psychic energy . . .

— JEAN PIERRE CHANGEUX, *Neuronal Man*

From Nervous Energy to Psychic Energy: Freud, or the Brain Diverted

THE PROBLEM OF ENDOGENOUS EXCITATION

Let us be clear: Freud never said that neurons are limited to "stimuli from outside." Very much to the contrary, from *Project for a Scientific Psychology* onward, he insists upon the existence of *endogenous excitation*, *internal* to the nervous system, an *excitation from inside*. He will never reconsider this point. However, he will soon go on to show that the nervous system, paradoxically, cannot respond to such excitation—which is to say that it is not in a position to regulate its own stimuli.

Freud will eventually identify this internal pressure with the *drive*, whose exercise, regulation, and channeling demands a specific space that, in a sense, obviates the nervous system and diverts a share of its excess energy.

This space is that of the psychic apparatus. Accordingly, the problem of the endogenous excitation of the nervous system is not fundamentally a neurological problem. This is the point of greatest concern in contemporary debates.

There are not, for Freud, two types of psychic energy: nervous energy and psychic energy. Instead, there is a relay, a differentiation, an economic complexification of the first type of energy by the second.

Let us recall the famous statement from the metapsychological papers according to which "the stimulus of the drive does not arise from the external world but from within the organism itself."[1] Or further: The drive "impinges not from without but from within the organism."[2] These affirmations echo analogous statements from *Project for a Scientific Psychology*:

> With an increasing complexity of the interior [of the organism], the nervous system receives stimuli from the somatic element itself—endogenous stimuli— which have equally to be discharged. . . . From these the organism cannot withdraw as it does from external stimuli; it cannot employ their Q for flight from stimulus.[3]

As we know, Freud distinguishes between *phi* neurons, which manage external stimuli, and *psi* neurons, whose task is to liquidate endogenous excitation.[4] In every case, however, Freud recognizes the existence of an "attack" that arises from inside the nervous system. At no moment does he seek to discover an energy that would be foreign to nervous energy, that would come from outside to manage the inside of the system.

How, then, does properly "psychic" energy—not to be confused with nervous energy—come upon the scene? It is invoked at the precise moment when Freud confronts the nervous system's inability to master its own stimuli, to satisfy the demands of these stimuli, precisely because they are *internal*. Freud will quickly abandon the hypothesis of the existence of two different types of neurons—permeable and impermeable—as a means of understanding and resolving the specific problem of endogenous excitation.

In Freud's metapsychological theory, the drive appears as a force that decidedly threatens the contact barriers, the facilitations, and the different "neuronal" systems for protecting against excessive quantities of excitation. The difficulty lies in the fact that this force is not exterior to the nervous system itself. On the contrary, it is the very manifestation of its inside. The drive appears as a force from inside that the inside of the system cannot manage. Freud declares:

> We see then how greatly the simple pattern of the physiological reflex is complicated by the introduction of the drive. External stimuli impose only the single task of withdrawing from them; this is accomplished by muscular movements, one of which eventually achieves that aim. . . . Stimuli coming from the drive, which originate from within the organism, cannot be dealt with by this mechanism. Thus they make far higher demands on the nervous system and cause it to undertake involved and interconnected activities by which the external world is so changed as to afford satisfaction to the internal source of stimulation.[5]

These "interconnected activities" reveal that the nervous system does not know how to manage itself; does not know how to manage itself as soon as it finds itself confronted with another dynamic than the primary logic of the reflex—that is, as soon as it finds itself confronted with an urgency, a pressure, a "constant force" *from inside*. Nervous energy is unable to propose or to produce a form of exteriorization or liquidation analogous to muscular discharge in order to rid itself of this internal thrust.

Psychic energy thus takes up where nervous energy leaves off, functioning, in a sense, as *the force of force*, which makes possible *what is impossible for the nervous system*: to find a means of appeasing or satisfying this constant, urgent, threatening internal excitation. It will be objected that the nervous system is perfectly well equipped to respond to internal stimuli. Doesn't pain, for example, constitute such a response? This, however, is precisely the point: in the case of the drive, there is no wound, no pain, and no localized cause of irritation. *It is the "inside" in its entirety that impinges.*

The problem thus appears with great clarity: the inside of the system (its economy) and the inside from which the excitation originates (the source of the drive) paradoxically become foreign to one another, as if they did not share the same interiority. The second inside, in fact, threatens to exceed the limits of the first even as they inhabit the same body, the same soul. The same inside is thus differentiated or divided from itself—the nervous system not being able to identify and hence to internalize the origin or the reason for this internal pressure—which results in a double energetic aim—a double aim that corresponds, in Freud, to *the doubling of the nervous system with a psychic apparatus.*

Once again, there are not two types of energy but rather a differentiated organization of energy that is implemented in order to address the problem of the *compatibility of the inside with itself*. The drive, Freud affirms, "obliges the nervous system to renounce its ideal intention of keeping off stimuli, for they maintain an incessant and unavoidable afflux of stimulation."[6] Psychic energy comes upon the scene precisely at the moment of this "renunciation."

The only way to engage the drive through its aim, which is to achieve satisfaction and to maintain excitation at the lowest possible level, is to divert the urgency of the drive from the inside, to deport the inside of the drive itself into *another inside* than that of the nervous system—another inside that thereby becomes, in a sense, external to the nervous system. This other inside is constituted by the psychic apparatus.

Psychic energy thus takes up where nervous energy leaves off in order to assume the task, in place of the brain, of responding to the unmasterable insistence of its self-irritation or of the inside convoking itself. Freud upholds a thesis that runs through his entire work: *the nervous system is deprived of any structure of auto-affection.*

THE BRAIN AS ELECTRICAL SYSTEM

What does this mean? Freud never contested the pertinence of the metaphor of the brain as electrical system that Breuer developed in *Studies on Hysteria*. In his contribution to this work, Breuer describes the brain as a pure and simple medium for the transmission of energy:

> We ought not to think of a cerebral path of conduction as resembling a telephone wire which is only excited electrically at the moment at which it has to function (that is, in the present context, when it has to transmit a signal). We ought to liken it to a telephone line through which there is a constant flow of galvanic current and which can no longer be excited if that current ceases. Or better, let us imagine a widely-ramified electrical system for lighting and the transmission of motor power; what is expected of this system is that simple establishment of a contact shall be able to set any lamp or machine in operation. To make this possible, so that everything shall be ready to work, there must be a certain tension present throughout the entire network of lines of conduction, and the dynamo engine must expend a given quantity of energy for this purpose. In just the same way there is a certain amount of excitation

present in the conductive paths of the brain when it is at rest but awake and prepared to work.[7]

At the same time, there is "an optimum for the height of intracerebral tonic excitation."[8] If this optimum is exceeded, the equivalent of a "short-circuit" is produced in the nervous system:

> I shall venture once more to recur to my comparison with an electrical lighting system. The tension in the network of lines of conduction in such a system has an optimum too. If this is exceeded its functioning may easily be impaired; for instance, the electric light filaments may be quickly burned through. I shall speak later of the damage done to the system itself through a breakdown of its insulation or through "short-circuiting."[9]

The brain, therefore, can only confront its own energetic excess through dysfunction. It is not equipped with any structure that would make it possible to dilute the excess by detouring—which is to say, differentiating—this energy. In other terms, *it cannot rely upon any apparatus of representation.*

DRIVE AND REPRESENTATION

Let us clarify the point. The "inside" from which the drive derives is not a substantial interiority, but rather a border zone, an undecidable limit "between the psychic and the somatic"; or, as Freud often said, between the "soul" (*Seele*) and the "body." The drive emerges from a certain excitation of the soul and the body as an ensemble. At the same time, it is inseparable from the way in which this ensemble is produced; that is, from a thrust that the brain—which remains indispensable to this ensemble—cannot manage to manage.

The "thrust" of the drive, despite the urgency of the pressure that it exercises, only manifests itself through representation or through delegation. It takes the time, at its origin, to double itself into mandating and mandated instances. The drive thus sends out its representatives in order to say that it cannot wait. It is this representative structure that qualifies and characterizes the particular relation between the somatic and the psychic at work in the drive.

Freud develops this argument in two apparently contradictory ways:

> Sometimes the drive itself is presented as the "psychical representative of the stimuli originating from within the organism and reaching the mind." At other

times the drive becomes part of the process of somatic excitation, in which case it is represented in the psyche by "instinctual representatives" which comprise two elements—the ideational representative and the quota of affect.[10]

In his metapsychological paper entitled "Repression," Freud affirms that an instinctual representative is "a representation or group of representations which is cathected with a definite quantum of psychical energy (libido or interest) coming from an drive," or, alternately, with a "quantum of affect."[11]

The "instinctual representative," therefore, doubles itself a second time, with a supplementary scission, into a representation or group of representations and a quantum of affect. Accordingly, Freud explains the way in which the force of the drive is destined to divide:

> A drive can never become an object of consciousness—only the representative that represents the drive can. Even in the unconscious, moreover, a drive cannot be represented otherwise than by a representation. If the drive did not attach itself to a representation or manifest itself as an affective state, we could know nothing about it.[12]

The fact that the drive is linked to an emissary—a representation or quantum of affect—thus constitutes its condition of possibility. Accordingly, the drive can always detach itself from the emissaries to which it might be linked; it can change emissaries in order to change its own "vicissitude."

The relation between the mind and the body only manifests itself within the limits of this disjunctive synthesis, this dissociative gathering, not of the body and mind themselves, but of the structure of their mutual representation or delegation. The extreme pressure that the drive exerts upon the nervous system, therefore, is both quantitative and qualitative: The thrust of the drive is at once a quantity of force and the division of the instances related to one another within force, the structural complication of force. This complication constitutes the inside that appears so foreign to the nervous system. Scission or separation becomes the only possible solution to the excess of endogenous irritation: division, stalling, and delay make it possible *to divert the pressure from inside without causing a "short-circuit."*

Freud insists, therefore, upon the fact that *the representative union of mind and body fails to be represented anatomically and organically within the nervous system.* Psychic energy detaches itself from nervous energy precisely in order to make possible this cloven representation, divided in its very unity, at the basis of all the vicissitudes of the drives: "reversal into its opposite, turning round upon the subject's own self, repression, sublimation."[13] Psychic energy is, in a certain sense, the rhetorical detour of nervous energy. Not able to discharge itself within the nervous system, endogenous excitation is diverted into roundabout paths, into turns comparable to tropes or figures of discourse. As Michel de Certeau quite rightly underscores, "the operations which order representation by articulation through the psychic system are in effect rhetorical: metaphors, metonymies, synecdoches, paranomasia, etc."[14]

At the same time, this rhetoric supplements the silence of the nervous system. Because no symbolic activity exists in the nervous system, psychic energy figures this very absence in a style that is inherently foreign to the brain, which is without initiative and does nothing but transmit energy or, as far as possible, maintain it at a constant level. *The unconscious is structured like a language only to the extent that the brain does not speak.*[15]

"Psychic" energy thus comes to sublate the absence of representative and symbolic power within the organization of the brain. The nervous system, whose first task is to master stimuli, cannot represent the relation of representation that originarily unites and disunites the psyche and the body. It cannot rely upon representation to elaborate "the organism's internal sources of excitation" which are themselves already structures of representation. The more the brain can affect itself, the more it is *not a psyche*.

SEXUALITY AND DIVISION

The partition of cerebrality and sexuality can be further clarified at this point in our analysis. Sexuality, it must be recalled, does not primarily designate sexual drive or sexual life, but rather a certain regime of events governed by a specific causality. However, the source of this causality lies in the division— at work within every drive, sexual or not (since, as Freud writes, all the drives are qualitatively the same)—between representation and affect.

The properly sexual regime for regulating events is organized around the division, the displacement, or the detour, which make possible the separate vicissitudes of the unity of the drive.[16] *Precisely that which cerebrality cannot manage is called "sexual"*: representation and substitution, delegation and separation, the distinction between representatives and affects. The sexual drive is, in a sense, reducible to the afflux of energy that presses forth and knocks at the door, demanding to be liquidated. But this liquidation is never simple or immediate: it subjects energy to detours, scissions, and divisions, in order to transform it into the ciphered message of *representance*.[17] Sexuality is the hermeneutic adventure of psychic energy. The exogenous event, when it occurs, will necessarily be separated from its own exteriority as it is folded into the internal adventure of sense.

It will come as no surprise that, for Freud, the "libido," which is initially defined as affect linked specifically to the sexual drive *stricto sensu*, ends up designating the mobility or the "rhetoric" of the quantum of affect in general. Much as there is sexuality in a "large" sense, there is also consequently a "large" sense of the word libido. The labile character of the libido, which makes it susceptible to infinite detours, becomes the dominant trait of psychic energy in its entirety. The treatment of the sexual drive becomes paradigmatic of all vicissitudes of the drive.

Lacan shows that the libido, far from being the mere dynamic manifestation of the sexual drive, designates, in fact, "an undifferentiated quantitative unit that is susceptible of entering into relations of equivalence."[18] Even if the libido is not the only energy at work within the psyche—far from it, as we will see in the second part of our study—it gives its name to every energetic transaction. "Hence," Lacan pursues, "one would talk of transformations, regressions, fixations, and sublimations of the libido, a single term which is conceived of quantitatively."[19] The function of the libido is thus to unify a *field*. Not simply the field of the different phases and structures of sexual development, but also, precisely, the "domain of psychoanalytical effects" in general,[20] the energetic tropes or tropisms that go beyond the organization of the nervous system.

The Redefinition of the Brain as Psyche

However, we must affirm as forcefully as possible: The Freudian conception of the brain as foreign to symbolic activity, as a pure material base, a pure support, without autonomy in the management of its own energetic impulses, is today entirely open to question. The revelation of the "emotional brain"—ready to endure stimuli originating from its own inside and thus capable of producing representations of this inside—plays a fundamental role in the elaboration of a new approach that assimilates brain and psyche, displacing the limits that stand between cerebral organization and psychical topology.

A NEUROBIOLOGICAL THEORY OF EMOTION: HOMEOSTASIS AND AFFECT

"Emotion" seems to be a much older word than "drive." In fact, "emotion" is not really a concept but rather a synthesis-word that ultimately designates something that neither "classical" psychoanalysis or neurology were capable of thinking: a dynamics (since, in "emotion," we hear movement) of the relation between the brain and the body, *the very movement of the psychosomatic totality,* made up of a singular body and a nervous system. There is a constant exchange of information between the two. Life regulates itself by informing itself about itself, and this self-information constitutes the elementary form of cerebral activity.

The origin of the dynamics of emotion lies precisely in this elementary activity. To begin with, emotion does not designate such and such an affect or passion, but rather a process at work in the regulation of life: there is a *pure vital emotion* without any object other than the "self"—the cerebral "self."

Preoccupied with the logic of the drive, psychoanalysis failed to notice *the process, at work within the homeostatic functioning of the nervous system itself, whereby the regulation of life gives rise to affects.* Far from being a mechanical process, comparable to the functioning of a generator, *homeostasis is an affective economy.*

The elementary activity of the nervous system is maintaining excitation at the lowest possible level conducive to survival, and this activity produces affects. The brain affects itself while regulating life. Accordingly, there is no "principle of inertia"—the name that Freud gives to the principle of constancy—without emotion; that is, without *auto-affection of the mechanism producing constancy.* "Curiously enough," Antonio Damasio writes, "emotions are part and parcel of the regulation we call homeostasis."[21] Whence a paradox that contains the stakes of our entire analysis: maintenance, constancy, inertia, and homeostasis are the paradoxical products of auto-excitation. In order to understand the "emotional brain," therefore, we must start from this paradox.

The emotions organize and coordinate cerebral activity. Whether it is a matter of primary emotions (sadness, joy, fear, surprise, disgust), secondary or "social" emotions (embarrassment, jealousy, guilt, pride), or what are called "background" emotions (well-being, malaise, repose, discouragement, etc.), the emotions are the elaborate prolongation of affective processes at work within homeostatic regulation. Within the brain, therefore, there are no regulatory mechanisms for adaptation to the outside that do not entail the emotive adaptation of the inside to itself. Every individual history begins there.

It would seem that psychoanalysis has remained blind to this cerebral auto-excitation. However, it will be immediately objected, doesn't Freud say that homeostasis, or the "principle of inertia," is regulated by the pleasure principle?

> When we further find that the activity of even the most highly developed mental apparatus is subject to the pleasure principle, i.e. is automatically regulated by feelings belonging to the pleasure-unpleasure series, we can hardly reject the manner in which the process of mastering stimuli takes place—certainly in the sense that unpleasurable feelings are connected with an increase and pleasurable feelings with a decrease of stimulus.[22]

Notwithstanding, it should be clear that the pleasure principle *does not give itself pleasure* in this sense that it does not affect itself. The mechanism of the pleasure principle remains inaccessible and insensitive to that of which it is the principle. This is why Damasio can uphold the strange affirmation that "pleasure is not an emotion."[23] Emotion is a reflexive structure through which vital regulation affects itself. "The status of life regulation is expressed in the form of affects."[24]

WHAT DOES ANATOMY TEACH US?

There are a limited number of sites in the brain that induce emotion. These sites occupy a zone that reaches from the brain stem to the upper brain. Excluding a part of the frontal lobe that is called the ventromedial prefrontal cortex, most of these sites are subcortical sites—sites that Freud considered to have no relation to the unconscious. The main subcortical sites are found in the region of the brain stem, the hypothalamus, and the basal telencephalon. The amygdala, or amygdaloid complex—an almond-shaped group of neurons situated in the temporal lobe in front of the hippo-campus—is also a subcortical site that plays a key role in the triggering of emotions. In particular, it forms the part of the limbic system involved in fear and aggression.

What anatomy shows us about the relation between the triggering and execution of emotions is that the distribution of emotional processes across several sites makes it possible for the brain to discipline and manage its internal sources of excitation *without being overwhelmed by them*. Indeed, these "sites" are not rigid and fixed but rather constitute functional systems. Damasio emphasizes that "none of these triggering sites produces an emotion by itself. For an emotion to occur the site must cause subsequent activity in other sites. . . . As with any other form of complex behavior, emotion results from the concerted participation of several sites within a brain system."[25] We might then conclude that *the foreignness of the inside to itself*, born of the intensity of excitation from internal sources, would be, in reality, *managed from within the nervous system* in a functional and interactive manner, thanks exclusively to the collaboration of several *cerebral* agencies. *The psychic detour imposed upon nervous excitation is no longer necessary.*

The brain itself thus deploys the rhetoric of its energetic excess. Which implies, in turn, that the activity as cerebral representation does exist and that the logic of affects at work in cerebrality should not be confused with the representation economy of the drive. This is the decisive point. Indeed, if the simultaneously synthetic and disjunctive structure of the psychosomatic relation is no longer situated outside the brain, if it no longer requires a detour, then *wouldn't this change the sense of the unconscious itself?* Accordingly, wouldn't it transform the vicissitudes of psychic energy that have been, until now, essentially conceived as libido?

THE CONCEPT OF THE CEREBRAL UNCONSCIOUS

This argument does not draw upon anatomy to postulate a brutal opposition between emotion and drive. Taking into account the anatomical localization of the circuits of emotion raises questions that this account alone does not suffice to resolve. Researchers such as Mark Solms, for example, have categorically refused to *localize* the unconscious. There is no question, Solms writes, of saying that "the unconscious is in the right hemisphere," nor that it corresponds to the emotion-inducing sites of the brain. If there is indeed a cerebral unconscious, linked to the emotional brain, it cannot be purely and simply assigned to a particular a "zone" or "region."

But the anatomical localization of the sites of emotion does remind us that the concept of the cerebral unconscious is linked to the brain's management of internal stimuli and to the auto-representational activity attached to it. This unconscious is constituted within the "core" where the originary intrigue of the individual's attachment to life is devised. This "core" corresponds to the neuronal elaboration of both a constant and ever-changing representation of the psychosomatic relation.

Accordingly, there exists *a form of the cerebral synthesis of difference*. This synthesis, which is the fruit of a representational activity, corresponds to a process of imaging: "Core consciousness occurs when the brain's representation devices generate an imaged, nonverbal account of how the organism's own state is affected by the organism's processing of an object."[26] This cartography of the relation between inside and outside reveals the "history" of the organism, "caught in the act of representing its own changing state as it goes about representing something else."[27]

The core, which is also called the "proto-self," the primitive form of identity, is thus constituted by constant interaction between the internal milieu and the external world. The state of the internal milieu, the viscera, and the musculoskeletal system (elementary homeostatic indices), produce continuous and dynamic representation through which life is constantly informing itself about itself. Instant after instant, the brain represents (to itself) the interaction between its internal state and external stimuli. The sources of internal stimuli are thus always identified and comprehended without having to be diverted:

> The proto-self is a coherent collection of neuronal patterns which map, moment by moment, the state of the physical structure of the organism in its many dimensions. This ceaselessly maintained first-order collection of neuronal patterns occurs not in one brain place but in many, at a multiplicity of levels, from the brain stem to the cerebral cortex, in structures that are interconnected by neuronal pathways. These structures are intimately involved in the process of regulating the state of the organism. The operations of acting on the organism and of sensing the state of the organism are closely tied.[28]

The mechanisms described in this passage constitute part of an ensemble of structures that simultaneously regulate and represent corporeal states. Within the brain, therefore, there is no regulation without representation.

This double economy defines cerebral identity as a constant synthesis of different states of relation between the body and the psyche.

THE AUTO-AFFECTION OF THE BRAIN

A question immediately arises: If we define the unconscious as nonconscious activity, don't we land once again in the trap that Freud famously denounced of confusing the unconscious (*Unbewusst*) and the nonconscious (*Bewusstlos*)? Don't we thereby entirely fail to grasp the signification of the *psychic* unconscious? Accordingly, as we move from psychoanalysis to neurology, are we dealing with the same concept of "representation"?

There is no doubt that, if we simply characterize the cerebral unconscious as the nonconscious place from which homeostatic processes are managed, we do indeed risk falling in this trap and adhering to a very insufficient, precritical, definition of the unconscious. However, this is not the case if we determine the "cerebral unconscious" as the *"cerebration" of affects*[29] (to borrow Marcel Gauchet's term)—that is, as an active and sui generis process of regulation. All the information that the brain offers (itself) about the internal state of the organism and about the relation between the organism and objects are accompanied by the production of affects. It is impossible to separate "information" and "sensorial modality": strictly speaking, *the brain feels itself informed*. The autorepresentative activity of the brain, ceaselessly mapped out within psychosomatic states, thus scrutinizes its own inside, translates it into images and affects itself with this activity, of which, we see, it is both sender and receiver. The "cerebral unconscious," then, designates less the entirety of nonconscious processes than *the auto-affection of the brain itself in its entirety*.

Homeostatic processes, the birth of the self, and the birth of object relations are bound together from the very beginning as one and the same phenomenon within the brain. The logic of cerebral auto-affection does not suppose the intervention of a supplementary energy that would have the status of the libido. The division between the self and the object is given before any narcissism and any sexual investment. Cerebral auto-affection is a logical sensuality that makes possible the attachment of life to itself which becomes the basis of all ulterior erotic investments.

Within the cerebral unconscious, homeostatic regulation puts in place, from the outset, the differentiated synthesis of self and object, survival and eroticism—without according the latter extraneuronal status. The psyche thus becomes the core that gathers, within the same energetic economy, the constant exigency of survival, the self's bond to itself, and the desire of the other.

How, precisely, do we understand the concept of the "auto-affection" of the brain? In philosophy, the notion of "auto-affection" traditionally designates the originary and paradoxical manner in which the subject experiences himself as self-identical by addressing himself as an other within the strange space of its "inner self" *(for intérieur)*. It constitutes a kind of primordial self-touching: the subject smells itself, speaks to itself, hears itself speaking, experiences the succession of its states of consciousness. It is this "contact" that produces the difference of the self from itself, without which, paradoxically, there would be no identity and no permanence. Auto-affection is the subject's originary ability to interpellate itself, to solicit itself, and to constitute itself as a subject within the double movement of identity and otherness to itself.

To speak of cerebral auto-affection, therefore, is to admit that the brain is capable of looking at itself, touching itself as it constitutes its own image. Homeostatic regulation has a specular structure; it operates as a kind of mirror within which the brain sees itself live.[30] Cerebral auto-affection, which designates the set of homeostatic processes, thus characterizes *the brain's capacity to experience the altering character of contact with itself.* Emotion plays a fundamental role within the constitution of this cerebral psyche: the brain affects itself—that is, modifies itself—within the constant flow of vital regulation. The stakes of neurobiological research consist in highlighting, on the basis of this elementary relation of the brain to itself and to the other, the idea of a cerebral identity that is not identical with subjective identity to the precise extent that it constitutes its unconscious.

What does this mean? The nature of cerebral auto-affection is different than the auto-affection of the subject as the philosophers have defined it. The elementary reflection that constitutes the cerebral psyche as such *does not reflect upon itself.* It does not redouble its specularity to the point of endowing it with the form of consciousness. *No one can feel his or her own brain; nor can he or she speak of it, hear it speak, nor hear himself or herself speak within it.* Cerebral auto-affection is necessarily and paradoxically accompanied by a blindness, *an inability of the subject to feel anything as far as it is concerned.* If the subject can "touch" itself, it is indeed thanks to the brain: the first contact with oneself constituted by homeostasis renders such auto-interpellation possible. At the same time, however, this originary solic-itation hides itself within the very thing that makes it possible. Within my inner self, my brain never appears. *The brain absents itself at the very site of its presence to self.* It is only accessible by means of cerebral imaging technology. And there is no possible subjectification of this type of objectification.

As I write these lines, I see myself write them, but this vision is only the distantly derived and thoroughly elaborated form of a primary auto-affection—cerebral auto-affection, constant but invisible, which forever keeps me from experiencing the wealth of energy that it contains and that makes it possible for me to write in the first place. *Cerebral auto-affection is the unconscious of subjectivity*.

Neurobiologists sometimes seem to recognize a certain proximity between the "proto-self" and the "ego" from Freud's second topic.[31] The ego, much like core consciousness, appears as a perceptive surface where internal excitations and external stimuli, coming from opposite directions, intersect. The eminent neurologist Jaak Panksepp goes so far as to define the "self" of the "proto-self" as a "Simple Ego-like Life Form." "This primal self," Mark Solms writes, "forms the foundational ego upon which all our more complex representations of our selves are built."[32] Freud's "ego" and the neurobiologist's "self" are both concepts of the border that mediates between the perception of internal states and the perception of external states. "The ego," Freud writes in *The Ego and the Id*, "is first and foremost a bodily ego; it is not merely a surface entity, but is itself the projection of a surface."[33] This description seems indeed to ascribe the representation of the body to a specific region of the brain that functions as the interface between internal sensations, perception, and motility.

In reality, the analogy does not go very far. Antonio Damasio categori-cally refuses to assimilate the Self to a "homunculus," the "little person" whom so many psychologists and neurologists—Freud included—con-ceived as the "inhabitant" within the ego. In *The Ego and the Id*, for example, Freud declares: "If we wish to find an anatomical analogy for it we can best identify it with the 'cortical homunculus' of the anatomists, which stands on its head in the cortex, sticks up its heels, faces backwards and, as we know, has its speech arena on the left-hand side."[34]

The homunculus corresponds to a figural representation of a part of the nervous system. There would thus be, in some sense, a subject within the subject, a little ego within the ego "that possessed the knowledge needed to interpret images formed within the brain."[35] "There is no homunculus involved," Damasio vigorously affirms:

> Nor is the proto-self to be confused with the rigid homunculus of old neurology. The proto-self does not occur in one place only, and it emerges dynamically and continuously out of multifarious interacting signals that span varied orders of the nervous system. Besides, the proto-self is not an interpreter of anything. It is a reference point at each point where it is.[36]

Of course, we know and can localize the cerebral structures necessary for the constitution of the proto-self, but this does not prevent the proto-self from remaining paradoxically unlocalizable, dynamic, and dispersed. The cerebral unconscious is not the "ego"; it is not an agency but rather a sequence, a constellation.

In this sense, the process of representation at work within cerebral auto-affection is very different from the process of drive representation analyzed earlier. Within the brain, affect does not detach from itself; it does not deprive itself of its own energy; it does not delegate itself; it does not meta-phorize itself. Nonetheless, it is also not the expression of a unity. Indeed, the "self," at its very core, is not gathered; its manifestation is fundamentally temporal: the self exists only insofar as it lasts and produces itself from instant to instant:

> The story contained in the images of core consciousness is not told by some clever homunculus. Nor is the story really told by you as a self because the core you is only born as the story is told, within the story itself. You exist as a mental being when primordial stories are being told, and only then; as long as primordial stories are being told, and only then. You are the music while the music lasts.[37]

Cerebral auto-affection is the biological, logical, and affective process by which finitude is constituted within the living core of subjectivity without ever being able to become the knowledge of a subject. The cerebral self represents itself without presenting itself.

The Temporal Brain and the Destructible Unconscious

It is thus possible to measure the distance that actually separates the neuronal unconscious from the so-called psychic unconscious. The first is bound up with the passage of time, while the second knows nothing about it: "The processes of the system *Ucs.* are timeless; i.e. they are not ordered temporally, are not altered by the passage of time at all. Reference to time is bound up, once again, with the work of the system *Cs.*"[38]

These assertions are linked to those in which Freud states that the unconscious is unfamiliar with death: "at bottom no one believes in his own death, or, to put the same thing in another way . . . in the unconscious everyone of us is convinced of his own immortality."[39] To the contrary, cerebral auto-affection is the incessant internal announcement and reminder of mortality. Damasio declares:

> We do not have a self sculpted in stone, and like stone, resistant to the ravages of time. Our sense of self is a state of the organism, the result of certain components operating in a certain manner and interacting in a certain way, within certain parameters. It is another construction, a vulnerable pattern of integrated operations whose consequence is to generate the mental representation of a living individual being. The entire biological edifice, from cells, tissues, and organs to systems and images, is held alive by the constant execution of construction plans, always on the brink of partial or complete collapse.[40]

There is an indissoluble link between the fact that the brain is capable of representing its own inside and the imaging of the deep destructibility of the neuronal configurations making up the proto-self. *The brain never behaves as though it were immortal.* There is nothing at work within it that would correspond to its immemorial and imperishable shadow. The cerebral unconscious is thus fundamentally a destructible unconscious and it "knows" itself to be so. This biologico-symbolic "knowledge" is the original experience of fragility as the absolute exposure to the accident.

The core of subjectivity, when attacked, is quite simply at risk of dissolving. Within the Self, at the most elementary level, there is no repression, nor disavowal, nor formation of substitutes in the face of danger. Faced with a threatening event, the Self has only one way out—its own loss.

The only possible subjective experience of cerebral auto-affection is that of the suffering that follows its damage or interruption.

Brain Wounds: From the Neurological Novel
to the Theater of Absence

> By what plausible means can destruction of a brain region change personality?
>
> — ANTONIO DAMASIO, *Descartes' Error*

On Fragility

Cerebral auto-affection is a process that becomes all the more fragile and all the more exposed to the extent that *the event of its destruction constitutes the only proof of its existence for the subject.* The importance of this auto-affection can be revealed only negatively, through an accident—a wound, damage, or trauma—that happens to interrupt or disrupt its functioning.

Contemporary neurology thus insists upon the necessity *to think anew the relation between the unconscious and destruction*, negativity, loss, and death. Writes Joseph LeDoux:

> Before we examine what holds the self together, let's consider how fragile a patch job it is. The bottom line is simple: functions depend on connections; break the connections, and you lose the functions. This is true of the function of a single system . . . as well as the interactions between systems.[1]

The presence of a cerebral lesion (such as skull trauma, stroke, or encephalitis) more or less seriously affects the processes at work within cerebral auto-affection such that the patient's personality is transformed to such a degree that it might never regain its lost form.

Indeed, we now know that every cerebral deficit has repercussions upon the sites in the brain that induce emotion, even if they have not been directly damaged. Accordingly, disturbances of cognitive function, such as aphasia or amnesia, are accompanied by disturbances of emotion. *Every cerebral illness or lesion affects the brain's auto-affection.*

That there is a link between emotion and cognition, between the emotional brain and the rational brain, is no longer open to doubt, since the elementary formation of these different instances are originally linked within the brain's homeostatic processes. Consciousness and emotion are not separable. If the one deteriorates, the other necessarily does as well. High-level cognitive processes—such as language, memory, reason, or attention—are not necessary for the constitution of the "proto-self." These functions are, however, structurally linked to emotional processes, and the selective reduction of emotion is just as damaging to rationality as excessive emotion. Damasio declares:

> In recent years both neuroscience and cognitive neuroscience have finally endorsed emotion. . . . Moreover, the presumed opposition between emotion and reason is no longer accepted without question. For example, work from my laboratory has shown that emotion is integral to the processes of reasoning and decision making, for worse and for better. . . . The findings come from the study of several individuals who were entirely rational in the way they ran their lives up to the time when, as a result of neurological damage in specific sites of their brains, they lost a certain class of emotions and, in a momentous parallel development, lost their ability to make rational decisions. . . . Their ability to tackle the logic of a problem remains intact. Nonetheless, many of their personal and social decisions are irrational. . . . I have suggested that the delicate mechanism of reasoning is no longer affected . . . by signals hailing from the neuronal machinery that underlies emotion.[2]

When emotional "signals" disappear, reason loses the vital link that unites it with cerebrality.

Almost all of the cerebral sites associated with emotions are located close to the median of the brain and occupy a strangely "central" position. Buried within the center of our heads, distributed throughout the folds of the brain, is a strange sort of interior helmet that we wear. We cannot feel it, but it constitutes the most sensitive point of our fragility. This absolutely vulnerable zone can be wounded at any moment, and damage to it may cause *a radical transformation of our identity.*

But it is important to understand—because this is fundamentally what is at stake within psychopathology today—that *the disruption of cerebral auto-affection does not put an end to psychic life.* Psychic life survives damage inflicted upon certain zones of the brain, in particular the emotional brain, even if this damage is extremely severe as in cases of aphasia, akinesia, various forms of epilepsy, or absence seizures.

With the exception of deep coma, no matter how severe such disruptions may be, they do not imply that the patients who endure them are deprived of psychic life and have thus become "vegetables" beyond the reach of psychotherapeutic treatment.

The break between contemporary psychopathology and classical psychoanalytic practice occurs at precisely this point. The nature of the changes in personality caused by brain damage excludes the possibility of interpreting them as forms of regression. Indeed, these changes do not allow patients to return to a previous state, to seek refuge in a past of any kind, or to find even the most precarious relief in the labyrinth of their psychic history. These transformations occur through destruction. When a psyche is shredded, it corresponds to the birth of a new, *unrecognizable* person. Such phenomena demand new forms of treatment that would no longer be based on the investigation of the past, the exploration of memory, or the reactivation of traces.

What people with brain damage have in common are changes in personality that are serious enough to lead their family and friends to conclude that they have metamorphosed into another person:

> Prior to the onset of their brain damage, the individuals so affected had shown no such impairments. Family and friends could sense a "before" and an "after," dating to the time of the neurologic injury.[3]

There is a postlesional plasticity that is not the plasticity of reconstruction but the default formation of a new identity with loss as its premise. This is why Damasio speaks of the "lesion method," which draws lessons from damage: "This approach, which is known as the lesion method, allows us to do for consciousness what we have long been doing for vision, language, or memory: investigate a breakdown of behavior, connect it to the breakdown of mental states (cognition), and connect both to a focal brain lesion. . . . A population of neurological patients gives us opportunities that observations in normals alone do not."[4] Lesional plasticity thus reveals a strange sculptural power that produces form through the annihilation of form.

To the extent that neurological patients have suffered, to different degrees, disturbances in their inductors of emotion, their new identities of neurological patients are characterized by disaffection or coolness. A bottomless absence. There is a manifest link between traumatic wounds and behavioral disaffection. To the extent that every trauma induces disturbances within the core of the "self," all posttraumatic changes of personality present such disaffection or desertion. *Psychoanalysis has never said anything about these "cool" psyches.* How to treat them? How to care for someone when that person, properly speaking, no longer exists?

A Few Psychopathological Cases

DISAFFECTION

"X." is a nameless patient evoked in the introduction to *Descartes' Error*:

> There was only one significant accompaniment to his decision-making failure: a marked alteration of the ability to experience feelings. Flawed reason and impaired feelings stood out together as the consequences of a specific brain lesion.[5]

X. serves as the paradigm for understanding how this "failure" is linked to certain lesions that irreversibly impair the ability to experience emotion:

> Neurological patients with damage to the amygdala cannot trigger those emotions [fear and anger] and as a result do not have the corresponding feelings either. The locks for fear and anger seem to be missing, at least for visual and auditory triggers operating under normal circumstances.[6]

This describes the case of David, who is suffering from a major lesion to both temporal lobes that resulted in damage to the hippocampus—the primary site of mnesic processes—which must be intact in order for the brain to remain open to new facts. His amygdala has also been damaged. "David, who has one of the most severe defects in learning and memory every recorded, cannot learn any new fact at all. For instance, he cannot learn any new physical appearance or sound or place or word."[7] In addition, he also proves unable to express emotion and seems indifferent to his surroundings. He feels neither anger nor anxiety and seems not to be conscious of or to care about his condition.

Elliot, who suffers from frontal lobe damage, was "thoroughly charming but emotionally contained. . . . He was cool, detached, unperturbed even by potentially embarrassing discussion of personal events."[8] A man with a lively intellect, endowed with an excellent memory for dates, names, and political events, a good father, and a good husband, Elliot had developed a meningioma that placed increasing pressure on his frontal lobes. After surgery, exactly like Phineas Gage, "Elliot was no longer Elliot." "The surgery was a success in every respect, and insofar as such tumors tend not to grow again, the outlook was excellent. What was to prove less felicitous was the turn in Elliot's personality."[9] When he is confronted with images designed to provoke strong emotions—buildings collapsing during earthquakes, houses burning, people wounded in bloody accidents or on the verge of drowning—Elliot flatly declares that he does not feel anything. The images cause no reaction whatsoever.

The emotional life of such patients is thus extremely impoverished. Most striking is their unfeeling manner of reasoning—a phenomenon that, according to neurologists, directly threatens their ability to *decide*, that is, to evaluate the different options in play when it comes to making a choice. Only the emotional apparatus makes it possible to lend weight to various solutions that call for a decision. If this apparatus remains mute, decision becomes a matter of indifference: Everything is just as good as everything else, so nothing is worth anything. The disturbance of cerebral auto-affection produces a sort of nihilism in the patient, an absolute indifference, a coolness that visibly annihilates all difference and all dimensionality.

ABSENCE SEIZURES

Pushed to the extreme, such perturbation can give rise to veritable absences or to suspensions of selfhood. Two groups of patients make it possible to study these phenomena of absence: patients who suffer from *epileptic automatism* and those with *akinetic muteness*.

Epileptic Automatism. "Absence seizures, Damasio explains, are one of the main varieties of epilepsy, in which consciousness is momentarily suspended along with emotion, attention, and adequate behavior."[10] These episodes of absence resemble a freeze frame. The patient undergoes a brutal interruption of his activity. He remains awake, does not fall, and experiences no convulsions, but he is no longer there. He looks at others in a state of "utter bewilderment or perhaps indifference."[11] He no longer knows who they are, who he is, or what he is doing, and he hardly looks at his surroundings. "Most frequently within seconds, more rarely within a few minutes, the automatism episode would come to an end and the patient would look bewildered, wherever he would be at that moment."[12] When he returns to himself, the patient retains no memory of what has just happened.

Akinetic Muteness. This formula designates a disturbance that manifests itself through loss of speech and an inability to make the least movement. Consciousness is seriously diminished, if not totally suspended, which, once again, entails the radical suspension of emotions. The case of L. is significant:

> The stroke suffered by this patient . . . produced damage to the internal and
> upper regions of the frontal lobe in both hemispheres. An area known as the
> cingulated cortex was damaged along with nearby regions. She had suddenly
> become motionless and speechless. . . . She would lie in bed, often with her eyes
> open but with a blank facial expression. . . . The term neutral helps convey the
> equanimity of her expression... She was there but not there.[13]

The patient no longer manifested any emotional reaction and seemed neither surprised nor unhappy to be in the hospital. "Again, emotion was missing."[14]

The patient T., for her part, suffered a cerebral hemorrhage that produced extensive damage in the frontal lobe of both hemispheres. She suddenly lost the ability to initiate movement or speech. She would usually lie in bed, her eyes wide open, with a blank facial expression. "I have often used the term 'neutral,'" Damasio affirms once again, "to convey the equanimity —or absence—of such an expression."[15] After recovering the ability to speak, she "was certain about not having felt anguished by the absence of communication. Nothing forced her not to speak her mind. Rather, as she recalled, "I really had nothing to say.' . . . To my eyes Mrs. T had been unemotional. To her experience, all the while, it appears she had had no feelings."[16]

Agnosia and Anosognosia. Agnosia designates the incapacity to remember the identity of a perceived object. Accordingly, the patient cannot recall the name or the identity of a familiar thing that is placed before his eyes. Nor can he recognize his own reflection in a mirror. Nonetheless, this deficiency does not affect core consciousness but the visual and auditory cortexes.[17] However, another neurological disorder, called *anosognosia*, results from damage to the somatosensory cortical network, which creates a phenomenon of absence due to a disturbance of the proto-self. Composed of the Greek words *nosos* (illness) and *gnosis* (knowledge), anosognosia designates the patient's inability to perceive himself as ill. When a patient left completely paralyzed on the left side of his body by a stroke is asked how he feels, his response is: "Fine."

This inability to recognize his own illness, which follows lesions only within the right hemisphere and from paralysis of the left side of the body, results from the loss of a very specific cognitive function. The patient no longer receives sensory information coming from the body or, at least, he no longer knows that he receives it. Once again, this phenomenon creates a strange emotional indifference:

> No less dramatic than the oblivion that anosognosic patients have regarding their sick limbs is the lack of concern they show for their overall situation, the lack of emotion they exhibit, the lack of feeling they report when questioned about it. The news that there was a major stroke . . . is usually received with equanimity, sometimes with gallows humor, but never with anguish or sadness, tears or anger, despair or panic.[18]

This disavowal of illness is not due to psychological causes but entirely to brain damage. Nor can the resulting disruption of self-image be explained in terms of a narcissistic disorder. The transformation of identity emerges from a sudden, isolated event, unrelated to other events that constitute an individual life story. Cerebrality thus designates the axiological principle that governs such accidents.

"A LANDMARK BY HINDSIGHT"

By way of introduction, I briefly evoked the emblematic case of Phineas Gage, which has become what Damasio calls a "landmark by hindsight" in neuropathology. Let us recall that, in the summer of 1848, Gage was the foreman of a construction crew building a railroad across Vermont when he suffered a severe head wound that caused damage to his prefrontal cortex.

He miraculously recovered from his wound within two months, but afterward, "he was no longer Gage."

This man, whose affective and social behavior displayed spectacular modifications thus became someone else in the wake of his accident. This case can be considered paradigmatic because it demonstrates that damage to cerebral auto-affection can severely impair the image of self and produce changes in personality; and, therefore, that it is not possible to take organic lesions into account without their psychic repercussions. But, just as significantly, it shows that the event—an accident or a lesion, along with its repercussions—cannot be woven into the thread of an individual's history. This thread, in fact, has been definitively cut. Gage's case makes clear that a certain regime of eventality—what Freud calls "internal" events, governed by the axiological principle of sexuality—is overwhelmed by the appearance of another regime of eventality. Within such a regime of events, *something as simple as an iron bar can cause a total metamorphosis of the personality*. This is the *cerebral* causal character of the accident without signification.

Once again, therefore, we are confronted with the question of how it might be possible to treat such psychic wounds if no meaning—that is, the category of meaning that pertains to damage according to Freud—can ever be ascribed to them. Coolness, neutrality, absence, and the state of being emotionally "flat" are the basic indexes of the meaninglessness of wounds that have the power to cause a metamorphosis which destroys individual history, that cannot be reintegrated into the normal course of a life or a destiny, and that, therefore, must be recognized as such even though it is impossible to categorize them as neurosis, psychosis, or, more vaguely, "madness."

Literary Forms of Neuropathology

Nonetheless, cases of brain damage can be written and narrated. Like psychoanalytic cases such as "Dora" or "The Wolfman," such cases have their own form of narrativity. Oliver Sacks was the first to think deeply about the style and function of writing proper to the neurological case history. In his preface to *The Man Who Mistook his Wife for a Hat*, Sacks shows that the tradition of clinical writing, which he traces back to Hippocrates, reaches its apex with psychoanalysis and then gradually declines during the second half of the twentieth century with the birth of "impersonal neurological science."[19] Indeed, for Sacks, it is necessary to weave the patient's coolness, indifference, and the disintegration of emotion into a narrative intrigue that must not be disaffected itself. This only seems to be a paradox for stylistic indifference is not an adequate response to subjective indifference. *Narrative work is a clinical gesture.*

It is thus necessary to make people with brain damage into *cases* in the strong sense—that is, into paradigms, into mirrors in which we learn to look at ourselves. The tradition of clinical tales, according to Sacks, must be endowed with new life and granted a new destiny. He recalls Luria's observation, "The power to describe, which was so common to the great nineteenth-century neurologists and psychiatrists, is almost gone now. . . . It must be revived."[20] Luria reacted to this loss by composing what he called "neurological novels," of which *The Man With a Shattered World* and *The Mind of a Mnemonist* are the most famous examples. The first of these two "novels" recounts the story that a man wounded in the war, Zasetsky, tells of his own mutilated life. Gravely wounded in 1943 by bomb fragments, he assumes the task of telling "the story of an instant that destroyed a whole life. The story of a bullet that penetrated a man's skull, damaged his brain, and shattered a world, leaving him irremediably dislocated."[21] The wound caused massive damage to the parieto-occipital region of the left hemisphere of the brain. Luria showed how all aspects of the patient's life were themselves fragmented by the bomb fragments. Zasetsky finds himself plunged into chaos. He had become amnesiac and struggled with aphasic disturbances. He can no longer see nor even imagine his right side. In the face of these and many other difficulties, he set to writing a journal of his injury. Sacks writes:

> Such an enterprise—picturing and at the same time anatomizing a man, the dream of a novelist and a scientist combined—was first realized by Freud: and Freud's magnificent case histories instantly spring to mind when one reads Luria. Luria's case histories, indeed, can only be compared to Freud's in their precision, their vitality, and their wealth and depth of detail (though, of course, they are also quite different, as neuropsychology is different from psychoanalysis). Both explore, fundamentally, the nature of man; both are new ways of thinking about human nature.[22]

Sacks affirms that his own practice of clinical writing was inspired by Luria's neurological novels. He compares his patients to characters in epic narratives: "heroes, victims, martyrs, warriors. . . . We may say that they are travelers to unimaginable lands—lands of which otherwise we should have no idea or conception."[23] Each case history in *The Man Who Mistook His Wife for a Hat* presents one of these travelers.

The problem of such case histories is how to do justice, in the very writing of the cases, to the rupture of narrativity that ultimately characterizes each one, to the destructive power of the plasticity that they each manifest. The "hero" of *The Man with a Shattered World* declares:

> I've become a stranger since I was wounded. . . . Everything that I learned or experienced in life has just dropped out of my mind and memory, vanished for good, leaving behind nothing but an atrocious brain ache.[24]

The question, indeed, is how to discover a mode of expression for such a "brain ache," a rhetoric proper to it, which would account for the total change in the personality of the sufferer. But what rhetoric could possibly account for the breakdown of connections, for destructive metamorphosis? And who would write the aphasic's novel? Who would write the story of losing all affect? *What mirror could reflect a brain?*

Neurological case histories must also borrow much from theater—in particular, from Beckett's theater, whose form might provide inspiration for the staging of neuropathological "cases." On many occasions, Damasio compares the disorientation of his patients to that of Winnie from Beckett's *Happy Days*, who is the incarnation of *wakefulness without consciousness.*[25] Indeed, it is quite possible that consciousness can be absent from the state of wakefulness: "Patients with some neurological conditions . . . are awake and yet lack what core consciousness would have added to their thought process: images of knowing centered on a self."[26] Winnie asks:

> What would I do, what could I do, all day long, I mean between the bell for waking and the bell for sleep? (*Pause.*) Simply gaze before me with compressed lips. (*Long pause while she does so. No more plucking.*) Not another word as long as I drew breath, nothing to break the silence of this place.[27]

The theater of absence is the privileged expression of affective impoverishment and destructive metamorphosis. Its rhetoric comprises figures of interruption, pauses, caesuras—the blank spaces that emerge when the network of connections is shredded or when the circulation of energy is paralyzed. This theater is what Gilles Deleuze calls the theater of exhausted identity. Such identity is the possible born after the exhaustion of all possibles.

There is no longer any possible. . . . Does he exhaust the possible because he is himself exhausted, or is he exhausted because he has exhausted the possible? He exhausts himself in exhausting the possible, and vice-versa. He exhausts that which, in the possible, is not realized. He has had done with the possible, beyond all tiredness, "for to end yet again."[28]

Couldn't these words, which describe the Beckettian character, the "exhausted," also perfectly well characterize the neurological patient?

This possibility of the exhaustion of the possible, which forms the identity of the exhausted, has nothing to do with the persistence of an anterior state (as in Freudian plasticity). For the exhausted, nothing persists—or rather, it is the nothing that persists: "One was tired by something, but one is exhausted by nothing."[29] In exhaustion, "one combines the set of variables of a situation, on condition that one renounce any order of preference, any organization in relation to a goal, any signification."[30] The theater of exhaustion gives voice to coolness as absence of sense, staging it as absence of persistence, of revivification, or of regression.

We are now in a position to distinguish the rhetoric and narrative technique of psychoanalytic case histories from those of neurological case histories. They are, in fact, incompatible. Whereas the psychoanalytic narrative of the resistance of the indestructible within destruction privileges the persistence of childhood, the past, and psychic destiny, which become recognizable through the disturbance that solicits them, the neurological narrative of the destruction of the resistance to destruction stages the exhausted but surviving resources of a psyche that no longer recognizes itself.

The writing of neurological suffering—the theater of absence or the novel of "brain ache"—raises the vertiginous question of *the psyche's survival of its own annihilation.*

THREE

Identity Without Precedent

I am a kind of newborn creature.

— ZASETSKY/LURIA, *The Man with a Shattered World*

The Impossibility of Turning Back

Has neurology today undertaken a thinking and writing of destruction more radical than psychoanalysis? As soon as one examines the literature of contemporary neuropathology, the question becomes inevitable. The case histories, in particular, make it possible to contest the Freudian definition of psychic plasticity.

THE PERSISTENCE OF THE PRIMITIVE IN FREUD

As I indicated in the Introduction, plasticity, for Freud, designates the imperishable character of psychic formations. The clearest articulation of this definition of plasticity appears in "Thoughts for the Times on War and Death":

It is otherwise with the development of the mind. Here one can describe the
state of affairs which has nothing to compare with it, only by saying that that in
this case every earlier stage of development persists alongside the later stage
which has arisen from it; here succession also involves co-existence, although it
is to the same materials that the whole series of transformations [*Veränderungen*]
has applied. The earlier mental state may not have manifested itself for years,
but none the less it is so far present that it may at any time again become the
form of expression of the forces in the mind, and indeed the only one, as
though all later development had been annulled or undone.[1]

A careful reading of these assertions shows that the imperishable nature of
psychic life does not apply to all mental developments—every life experience or every event—but only to the fundamental psychic form, the initial
form that subsists throughout these developments even as it undergoes
transformation.

Plasticity must then be understood as a form's ability to be deformed
without dissolving and thereby to persist throughout its various mutations,
to resist modification, and to be always liable to emerge anew in its initial
state. It is precisely the series of transformations that can always "be
annulled" so that this "unique form" can reappear. Precisely and paradoxically, plasticity characterizes both the lability and the permanence of this
form.

This extraordinary plasticity of mental developments [*diese ausserordentliche
Plaztizität der seelischen Entwicklungen*] is not unrestricted as regards direction; it
may be described as a special capacity for involution—for regression—since it
may well happen that a later and higher stage of development, once abandoned,
cannot be reached again. But the primitive stages can always be re-established;
the primitive mind is, in the fullest meaning of the word, imperishable.[2]

As we can see, this persistent form is the primitive. The notion of primitivity thus gives content to the concept of the initial form of psychic life. The
primitive, and not what descends from it, is imperishable.

The primitive, for Freud, has two meanings, each rigorously articulated
in relation to the other. On one hand, the primitive is the "savage,"
also called "prehistoric man," who subsists within each of us. On the other
hand, as in the preceding passage, the primitive designates the "primitive
psyche"—that is, both the general psychic form in which the savage
survives within us and the particular psychic style of this survival: the constitutively unique character of the individual's *childhood*. Psychic plasticity

designates the combined persistence within us of prehistoric man and the child, the always open possibility of their imminent return.

The shadow of this return most often takes the form of a threat because *the revival of the primitive is what defines mental illness*. Indeed, Freud affirms that psychic disturbances always bear witness to the possibility of such resurgence:

> What are called mental diseases inevitably produce an impression in the layman that intellectual and mental life have been destroyed. In reality, the destruction only applies to later acquisitions and developments. The essence of mental disease lies in a return to earlier states of affective life and of functioning [*das Wesen der Geisteskrankheit besteht in der Rückkehr zu früheren Zustanden des Affektlebens und der Funktion*]. An excellent example of the plasticity of mental life is afforded by the state of sleep, which is our goal every night. Since we have learnt to interpret even absurd and confused dreams, we know that whenever we go to sleep we throw off our hard-won morality like a garment, and put it on again the next morning.[3]

Freud thus underscores two fundamental characteristics of psychopathologies: They always entail both regression and destruction, and they only destroy that which stands in the way of regression. Destruction only bears upon the "later acquisitions and developments" that Freud compares to a garment or an envelope. These superstructures are thus designed to cover the essential—the nature that breaks through our "hard-won morality"— the nudity of the primitive psychic stratum, which becomes the aim of regression. *Destruction is merely the most effective manner of uncovering or revealing the indestructible.*

NEUROLOGICAL DISEASE AND DAMAGE WITHOUT REGRESSION

Contrary to Freud's affirmations, however, it seems that the type of brain damage that we have been examining bears witness to *the impossibility of regression*, to a *point of no return*. In such cases, the core of the psyche is used up, exhausted, or destroyed. The fact that "what are called mental diseases inevitably produce an impression . . . that intellectual and mental life have been destroyed," is not, or no longer, the result of a supposed layman's error but has been definitively established by *neurological diagnosis*. When cerebral auto-affection is damaged, the Freudian definition of plasticity as the permanence of the primitive psychic stratum loses its relevance. When a lesion is present, there is, of course, a compensatory plasticity that attempts to supplement the damaged functioning; but, in cases of severe and irreversible cerebral pathology, this plasticity gives way to *another* plasticity. This is the plasticity that we have called destructive plasticity, the ability to create an identity *through loss of past identity*. An identity without childhood.

Cerebral pathologies bear witness to the psyche's ability to continue to live after its "initial state" has been destroyed, to survive the dislocation of its history. Despite its name, such survival does not appear as sublation or redemption. It has nothing to do with salvation or resurrection. On the contrary, it most often manifests itself as a strange way of being absent from life, disaffection of the ability to live and to die, unfolding of life without life to unfold. Winnie's life.

The loss of certain faculties can cause regressive behavior, but this turn is not exactly the return to a previous state because it does not regain anything familiar. Where it returns, there is nothing to find. The past is no longer a refuge.

In most patients with advanced Alzheimer's disease, the deterioration of consciousness is analogous to the cases of akinetic muteness described earlier:

> The decline first affects extended consciousness by narrowing its scope and progressing to the point in which virtually all semblance of autobiographical self disappears. Eventually, it is the turn of core consciousness to be diminished to a degree in which even the simple sense of self is no longer present. Wakefulness is maintained and patients respond to people and objects in an elementary fashion—a look or a touch, the holding of an object—but there is no sign that the responses issue from real knowing. In a matter of a few seconds, the continuity of the patient's attention is disrupted, and the lack of overall purpose becomes evident.[4]

There is wakefulness, there is awakening, but all signs of emotion, positive or negative, have been extinguished. The disease begins with cognitive disturbances, perceptive dysfunction, followed by the irremediable degradation of memory functions, along with modifications of personality and behavior, and, finally, alteration of cerebral auto-affection. The result is paradoxical behavior that is a combination of aggressivity (the patient is often irritable or agitated) and apathy (loss of initiative, dulled affectivity, disinterest in family and friends). In certain respects, such behavior looks like a return to childhood. The psychiatrist Barry Rosenberg even developed a theory of what he called "retrogenesis" in order to account for the process by which neurodegenerative mechanisms remove the latest psychic acquisitions first and then follow the inverse order in which they were achieved in normal development.[5] Nonetheless, this thesis remains controversial. Harry Cayton, president of the British branch of the Alzheimer's society, declares:

> Firstly, I have not managed to find many arguments in the scientific literature to support this theory. Secondly, if it is true that there is a similarity between certain affective and functional behaviors in people suffering from dementia and children, the underlying neurological processes are totally different.[6]

The exhausted brain is not a child.

Even if Alzheimer's patients seem to "fall back into childhood," it would still be possible to affirm that they return to *a childhood that is not their own*, to a childhood that is only a concept of childhood, that consists in a set of stereotypical gestures and postures that pertain to everyone's childhood and thus to no one's childhood. A childhood without a child to live it. John Bayley, the husband of Iris Murdoch, wrote an admirable account of the process whereby Alzheimer's disease engulfed the great writer. He evokes, for example, the mornings that Iris spent watching *Teletubbies*, a show intended for little children that she was especially fond of. Bayley judiciously notes that the writer had become *childish* but not *a child*. Childish but not the child that she had been.[7]

The Determining Character of "Modifications of Connections" and "Disconnections"

IS THERE A "PRIMITIVE" BRAIN?

It would seem that there is no neurological definition of what Freud called the "primitive psyche." What would be the meaning of the "primitive" within the economy of cerebrality? The reference to prehistoric man or to "savages" is not fundamental in neurology. Reflections upon the process of humanization and the development of the cortex in relation to the evolution of technology and the use of tools does not, properly speaking, belong to the field of neurobiology. There are no important studies that compare the organization of the primitive brain and the organization of the "developed" brain.

The motif of primitivity certainly enters into neurological studies that examine the genesis of the nervous system. Scientists distinguish, for example, between the "old" brain and the "young" brain. Certain regions of the brain, the brain stem, the hypothalamus, the base of the telencephalon, and very probably the amygdala and the cingulate cortex are evolutionarily old.[8] These regions, which are present in many species, control fundamental vital processes without relying upon cognitive processes.[9] Other regions, such as the neocortex, are evolutionarily modern and they preside over the processes that form mental images and intentional behavior. Accordingly, these are the parts of the brain that are most directly shaped by experience.

Nonetheless, the role of the old brain is not limited to homeostatic regulation; it also plays a role in the "development and adult activity of the evolutionarily modern structures of the brain."[10] The older circuits necessarily interfere "with more modern and more plastic circuits concerned with representing our acquired experiences."[11] These experiences are thus modulated by a set of primitive demands linked to survival. Accordingly, genetically determined bioregulatory circuits are informed by what happens in the most modern regions and constantly react to such events. This mutual influence is, in turn, mediated in large part by the "modulator" neurons located in the brain stem and the base of the telencephalon. The modulator neurons are then influenced by the organism's interactions with the world: "The unpredictable profile of experiences of each individual does have a say in circuit design, both directly and indirectly, via the reaction it sets off in the innate circuitries, and the consequences that such reactions have in the overall process of circuit shaping."[12]

However, when damage occurs, there is no dissociation between the old and the modern brain. The former does not resist the destruction of the latter as its "imperishable" core. Many of the inductors of emotion are located in the "old" brain (notably in the hypothalamus and amygdala). When these inductors are damaged, cerebral auto-affection is altered. The interaction of the old and the recent forbids the conclusion that there is a place in the brain where indelible memories are stored that is resistant to the effacement of traces. Cases of severe brain lesions show that *no stratum of cerebral organization is sheltered from destruction.*

It is thus important to modify both the paradigm of Freudian plasticity as the relationship between destruction and the indestructible and the Freudian definition of psychic disturbance. Brain damage does not confront us with regressive pathological processes that would progressively destroy the recent strata of the psyche in order to lay bare its primitive core. Within cerebral economy, there is no permanent form that can be transformed without being shattered. The pathological modification of cerebral connections does bring about changes of form but these changes utterly efface the previous form. Therefore, the paradigm of transformation of a form that remains the same must be displaced by that of transformation that creates a new form as it sweeps away the original. Within this new paradigm, psychic disturbance is no longer due to revival of past forms but rather to a forgetting of form. This is why neurobiology opposes models of the modification of connection or disconnection to the model of regression. Cerebrality, the evental regime of the accident thus characterizes the mode of production of such changes without memory, which are characteristic of interrupted auto-affection.

The complexity of the concept of a cerebral unconscious inheres within the relationship of the cerebral psyche to its own destruction. The idea of destructible psychic life gives body to the threat of irreversible damage—a threat that Freud does not recognize—that weighs upon unconscious functions. The fact that the entire brain—both "old" and "modern" regions—is exposed to lesional events, that mere accidents or raw shock can have decisive psychic repercussions that manifest themselves primarily in the form of changes in personality, all of these phenomena thoroughly transform the traditional psychopathological concept of causality and call for new etiological principles.

These new principles must take into account the significant—which is to say, psychic—impact of events without signification. Today, such impact is predominantly characterized in terms of the *modification of connections* and *neuronal disconnection*. LeDoux declares, "In general, there is a growing interest in the idea that alterations in synaptic connectivity in neuronal circuits, rather than just levels of neurotransmitters or receptors, are important."[13] Schizophrenia, for example, is accompanied by "alterations of function in specific brain regions and circuits rather than global changes in the level of monoamines."[14] The true causes of illness now appear to be the variations in the size or volume of certain brain regions that result from such modifications. Transformations or ruptures in form are generally what determine the extent to which the event of raw damage impacts the brain.[15]

Modifications of connections or disconnections are the operators of the profound upheaval within cerebral organization, beginning with the personality. Accordingly, one often speaks of "syndromes of disconnection" to characterize what happens when the communication between different regions of the brain is interrupted. However, "if the mental trilogy breaks down" because of ruptured connections, "the self is likely to begin to disintegrate and mental health to deteriorate. When thoughts are radically dissociated from emotions and motivations, as in schizophrenia, personality can, in fact, change drastically."[16] There is thus a clear link between the transformation, cutting, or interruption of connections and the metamorphosis of identity. "When connections change," LeDoux continues, "personality, too, can change. . . . When emotions run wild, as in anxiety disorders or depression, a person is longer the person he or she once was."[17]

WHEN THE PSYCHE NO LONGER DREAMS

The most illustrative example that Freud provides of his conception of the plasticity of psychic developments is the dream:

> An excellent example of the plasticity of mental life is afforded by the state of sleep, which is our goal every night. Since we have learned to interpret even absurd and confused dreams, we know that whenever we go to sleep we throw off our hard-won morality like a garment, and put it on again next morning.[18]

Today, however, we know that certain cerebral disconnections affect—to the point of destroying it—the capacity to dream. Mark Solms, in his book *The Brain and the Inner World*, explains the set of cerebral mechanisms at work in the process of dreaming. Three visual cortexes, situated under the occipital lobes, are directly involved in this process. The first cortex, connected to the retina, is called the "primary visual cortex." Next to it is an intermediary zone, which plays a central role in the visual recognition of color, form, and movement. Finally, the third cortex corresponds to a more complex type of visual activity in which vision "is dependent upon several other sensory modalities."[19] This zone of the brain "is involved in arithmetic, writing, constructional operations, and spatial attention."[20]

Damage to the first zone causes blindness in waking life, but the capacity to see in dreams remains intact. Damage to the second zone, however, disturbs vision both in waking life and in dreams: Patients no longer dream in color, the images in their dreams become static, and they lose the ability to recognize the faces that they see in dreams. Finally, damage to the third zone results in a "complete loss of dreaming."[21]

Freud considered dreaming to be the unconscious expression of the primitive psyche. But this example clearly shows that, although it happens every night, although it persists beneath all the disguises that waking life imposes upon it, dreaming cannot resist the destruction of the cerebral zones that make it possible. Is the psychoanalyst, then, capable of apprehending a psyche that has been deprived of the ability to dream *while remaining a psyche?*

What would Freud say, today, about emotional absence, the breakdown of symbolization, the deprivation of the capacity to dream? How would he understand the possibility of losing the mind's treasury of images, the primitive core, the child, the "savage"?

Might it not be necessary, today, to insist upon a "breach opened" within the theory of psychoanalysis by "those patients who rightly lament that they have no self, no 'me,' no individuality"?[22] Might it not be necessary to examine the theoretical, clinical, and political signification of these "catastrophic melodies" that haunt the diseased cerebral psyche? How, finally, could we not take into account another sense of plasticity than that which supposes the imperishable character of psychic acquisitions?

It is incontestable that Freud never said anything about a postlesional destructive plasticity that would create an identity without precedent. In *Civilization and Its Discontents*, of course, he does admit that excessively severe cerebral lesions interfere with the possibility of regression and reach the "primitive psyche":

> The assumption that everything that is past is preserved holds good in mental life only on condition that the organ of the mind has remained intact and that its tissues have not been damaged by trauma or inflammation.[23]

But precisely when a cerebral lesion is present, when "the organ of psychic life" is damaged, psychic life itself comes to a halt. What succeeds this psychic life is a vegetative state that apparently no longer pertains either to the competence or to the jurisdiction of psychoanalysis. For Freud, the disappearance of the primitive is tantamount to psychic death. Today, however, psychopathology must contend with the redoubtable fact that this supposed "death" is in reality a form of life.

FOUR

Psychoanalytic Objection: Can There Be Destruction
Without a Drive of Destruction?

Indeed, it may, as we know, be doubted whether any psychical structure can be the victim of total destruction.

— SIGMUND FREUD, *Constructions in Analysis*

Before continuing to examine psychic destruction and its creative plasticity, we must hasten to address the virulent critique that Freud would not have failed to mount against the preceding developments: Is it really possible to take into account a destruction of the psyche caused only by exterior events? Must we not always relate this destruction, in one way or another, to an internal power of annihilation that is immanent to the psyche itself? Isn't the way in which the psyche reacts to accidents conditioned by the self-destructive tendency at work within each individual? Isn't there an internal negativity that always doubles any fortuitous encounter with negativity? In a word, *can one really think destruction without a specific drive of destruction?*

It would appear necessary, at this point in our analysis, to test the hypothesis of a *plastic formation through destruction* against the Freudian hypothesis of the death drive. If cerebrality characterizes the relation of the psyche to emergent accidents and to death, it cannot, for this very reason, merely designate a regime of external events that threaten to interrupt cerebral auto-affection *without the latter's taking part in this interruption.* How does one think destruction without admitting that it *works*—much like a vicissitude, the vicissitude of a *drive*—within the psyche itself?

Indeed, it does not suffice to insist upon the finitude and fragility of the cerebral core; it is also necessary to consider the possibility that there is a *process of death* within the brain. Positing that the unconscious knows neither time nor death, Freud strangely shows that the psyche has a relation to destruction that is more radical than the biological knowledge of mortality.

The Persistence of the Trace

A first objection takes issue with the hypothesis of destructive plasticity and threatens to undermine its validity. Such plasticity, I have argued, has the power to form identity through destruction—thus making possible the emergence of a psyche that has vacated itself, its past, and its "precedents." In this sense, such plasticity has the power of creation *ex nihilo*, since it begins with the annihilation of an initial identity.

Such an analysis necessarily presupposes a fundamental distinction between the "normal brain"—with its positive plasticity of neuronal modulation, the economy of its affects, and the regulative activity of its auto-affection—and the damaged brain, with its negative plasticity, which causes an absolute metamorphosis of the subject through sheer accident. There seems to be no link between these two brains. However—herein lies the heart of the objection—can one seriously postulate such an absence of linkage or relation between the nonpathological functioning of the brain and its dysfunction? Within the ordinary processes of cerebral auto-affection, isn't there a secret propaedeutic that anticipates the metamorphosis of damaged identity?

This objection has three parts. First of all, how can one deny, even in cases of very serious damage, that something like a psychic structure or profile remains intact? How can one deny that a style of being endures despite the alterations and disturbances that it undergoes? Even if a subject no longer recognizes us, don't we always recognize him within his very metamorphosis? Is there really such a thing as an *unrecognizable psyche*? And, consequently, is there really a *clinic of the unrecognizable*? Doesn't every therapy, in one manner or other, start with the vestiges or the ruins of identity rather than with the metamorphic effects of its deracination?

It would be possible to provide ample testimony that even a very diminished person remains fundamentally who they always were. There is much proof of this ontological persistence of identity. Let us take the case that Oliver Sacks calls "The Lost Mariner." Jimmie suffers from Korsakov's syndrome, which entails a profound and irreversible loss of memory. This pathology is sometimes also called "transient global amnesia" (TGA). Jimmie, Sacks writes, "both was and wasn't aware of this deep, tragic loss in himself, loss *of* himself."[1] He had the very strong feeling of "something missing" but did not know what precisely it was and, for this reason, displayed a strange and profound indifference to his own "disappearance" ("Are you miserable?"—"I can't say that I am." "Do you enjoy life?"—"I can't say I do."[2]). At a certain point, Sacks wondered whether it would be valid to conclude that Jimmie had lost his soul: "Was it possible that he had really been 'desouled' by a disease?"[3]

Sacks asked the Sisters at the hospital: "Do you think he *has* a soul?"[4] Shocked by the question, they advise the doctor to observe two surprising aspects of Jimmie's behavior: his attitude toward his brother and his conduct in chapel.

Even though he was incapable of recognizing anyone, Jimmie was always very moved when his brother came to visit. He knew perfectly well who he was, even if he looked "old." "These meetings are deeply emotional and moving to observe—the only truly emotional meetings Jimmie has."[5] In chapel, then, Jimmie demonstrated an attention and concentration that he was totally incapable of in other situations. He was "absorbed in an act, an act of his whole being."[6] And Sacks concludes: "Perhaps there is a philosophical as well as a clinical lesson here: that in Korsakov's, or dementia, or other such catastrophes, however great the organic damage and internal dissolution, there remains the undiminished possibility of reintegration by art, by communion, by touching the human spirit: and this can be preserved in what seems at first a hopeless state of neurological devastation."[7]

The analysis of this case shows clearly that something remains, something on the order of a secret core of identity that resists the ordeal of trauma.

Between one neuronal plasticity and the other, positive and negative, creative and destructive, is there not, precisely, a plastic relation, a supple link? And doesn't this prove that, regardless of the apparent metamorphosis, the victim of brain damage or trauma has not entirely become someone else?

The second objection, which follows from the first, can be formulated in these terms: every trauma is a trace. Like any impression, it imprints the psyche with its seal, inscribes itself therein. The external event, as brutal and as senseless as it may be, always ends up making sense as a mark, writing. Boris Cyrulnik declares that, if trauma "alters the wounded person's behavior and emotions and often the deep regions of his brain responsible for emotions and memory," this does not prevent "the trace of the traumatic event . . . from living inside the psyche like a heavy crypt."[8] The traumatic event does entail a "biological metamorphosis," but the secret of this metamorphosis—the secret that would explain its triggering—remains, for its part, preserved within the intimacy of the originary psychic identity.

If it is true that the most widespread understanding of cerebral plasticity concerns the ability of neurons to change form in response to the subject's environment and experiences; if it is true that experience plays a fundamental role in the changes that affect the size and volume of synaptic connections, then why does the subject's experience come to an end as soon as the catastrophe happens? To what extent is the traumatic event also inscribed within the register of experience? How would it be possible to deny the relation between experience and accident?

It seems that we must acknowledge with Freud that, no matter the extent of "psychic modification," the metamorphosis of identity is never total; or that every modification is, in fact, "a change involving the same material and occurring in the same locality"[9]; or, yet again, that "portions of the earlier organization always persist alongside the more recent one and even in normal development the transformation is never complete."[10]

If my previous assertions concerning the profound metamorphic force of brain damage and the perishable character of the "primitive psyche" are viable, then it is extremely important to clarify the relation between formation and destruction within the brain. In other words, it would be necessary to consider that, in order to think the work of negative plasticity—that is, the evacuation of identity, absence from self, or absence to oneself—one must also postulate the existence of an internal, endogenous, process of destruction that responds to the traumatic stimulus and welcomes it, in a sense, facilitating its work of annihilation. Doesn't the possibility of indifference, coolness, and detachment always already mount a silent and originary attack against cerebrality even in the absence of any lesion, damage, or accident?

Cerebral auto-affection corresponds to finitude's auto-annunciation: the brain knows itself and calls itself fragile, mortal. But is this auto-annunciation of mortality pure of destructive tendencies itself? Couldn't one conclude, to the contrary, that postlesional behavior is the fulfillment of an internal dynamic that the lesion or accident can certainly precipitate or *trigger* but not entirely *cause*?

It would thus be necessary to postulate that there is a *conjunction* or a link between the work of cerebral auto-affection—which constitutes a *continuous* annunciation of finitude—and the event that comes to interrupt this very continuity, the traumatic intrusion that kills psychic identity. *How would it be possible to assert that there isn't anything within the homeostatic circuit of cerebral auto-affection that works toward the destruction of the system?* I have insisted upon the fact that cerebral economy is a representational structure, a mirroring structure. Couldn't it also be a machine of destruction? Of self-destruction or destruction of the other? Can the *initiative* of destruction be ascribed entirely to trauma?

A Neuronal Death Drive?

The attempt to negotiate between cerebrality and the economy of the death drive is thus inevitable. The death drive, in Freud, presents two essential characteristics that might seem contradictory. On one hand, the death drive embodies a tendency toward restoration. Present within every living substance, it manifests itself as the uncontrollable movement whereby life returns to an inanimate state. The "task" of the death drive "is to lead organic life back into the inanimate state."[11] It is inherent within the movement of life itself. The death drive is a set of tendencies "inherent in living substance towards restoring an earlier state of things; that is to say, they would be historically determined and of a conservative nature and, as it were, the expression of an inertia or elasticity present in what is organic."[12]

On the other hand, the death drive manifests itself as a tendency toward destruction, of which the most obvious manifestation is aggressivity, either directed against others or against the subject himself. However, this tendency toward destruction does not contradict the tendency toward restoration. It only ever results from a diversion of the latter toward objects in the external world. For Freud, we must suppose that "some portion of self-destructiveness remains within, whatever the circumstances until at last it succeeds in killing the individual. . . . Thus it may in general be suspected that the *individual* dies of his internal conflicts."[13]

Shouldn't we conclude that *cerebral lesions participate within these internal conflicts*, aggravate them, or precipitate them, and thus that the drive has final jurisdiction over neuropathology? How would it be possible to prove that indifference or coolness are not, ultimately, paradoxical manifestations of aggressivity or the return of life toward an inorganic state which, even for Freud, *has nothing to do with the return to childhood*?

Isn't the death drive inscribed immediately within the homeostatic movement itself? Lacan is quite right to assert that, "in man, there is already a crack, a profound perturbation of the regulation of life. That's the importance of the notion introduced by Freud of the death drive."[14] This crack is precisely *the crack of homeostasis*.

Indeed, the principal function of homeostasis[15] is restitution. It seeks to maintain energetic tension at the lowest possible level. The nervous system is constantly searching for this position of equilibrium that is tantamount to the survival of the organism. However, as Lacan remarks, "the minimum tension can mean one of two things, all biologists will agree, according to whether it is a matter of the minimum given a certain definition of the equilibrium of the system, or of the minimum purely and simply, that is to say, with respect to the living being, death."[16] The crack lies precisely in the fact that *homeostasis signifies both equilibrium and death*.

Freud is even led to give each of these twin principles distinct names: The first is called the *principle of constancy*, while the second is called the *Nirvana principle*.[17]

> The dominating tendency of mental life, and perhaps of nervous life in general, is the effort to reduce, to keep constant or to remove internal tension due to stimuli (the "Nirvana principle," to borrow a term from Barbara Low) . . . and our recognition of that fact is one of the strongest reasons for believing in the existence of the death drives.[18]

Homeostatic equilibrium does indeed demand the "leveling of chemical tension" that is death. This is why the principle of constancy can be defined as a process that both seeks equilibrium and leads the individual toward his own end.[19] Accordingly, if cerebral auto-affection is inseparable from the originary homeostatic regulation of the nervous system, it must also be traversed by this crack that, even as it creates the distinction between equilibrium and death, makes it impossible to extricate them from one another.

Freud declares: "If we are to take as a truth that knows no exception that everything living dies for internal reasons—becomes inorganic once again—then we shall be compelled to say that '*the aim of all life is death*' and, looking backwards, that '*inanimate things existed before living ones.*'"[20] "Gage is no longer Gage." "Elliot is no longer Elliot." Shouldn't we then consider these cases of metamorphosis as examples of life's self-destructive regression toward the inanimate, prefigurations of a return to inert matter, which the living being strives to regain at any cost? Doesn't trauma create a certain *ex-centricity* with respect to the immanence of destruction? In which case, doesn't neurology turn out to be more than the *ex-centric other* of psychoanalysis?

PART TWO

The Neutralization of Cerebrality

Introduction: Freud and Preexisting Fault Lines

All neurotics are malingerers; they simulate without knowing it, and this is their sickness. We have to keep in mind that there is a big difference between conscious refusal and unconscious refusal. The conscious and the unconscious are always conjoined in an individual, however, and if I confront a neurotic who claims and believes that he is organically ill with a statement that he is not, he will be offended, because it is partially true.

— SIGMUND FREUD, cited by Kurt Eissler in *Freud as an Expert Witness*

It is unquestionably difficult to break through the heavy wall of the Freudian theory of psychic events. More solid than the "protective shield" that it requires, it seems that this theory never admits anything from outside without filtering it through the endogenous processes of psychic life. Any conclusions drawn from the examination of neuropathological cases—with their total metamorphoses of identity, destruction of the primitive layers of the psyche, profound alteration of auto-affection, and irremediable annihilation—come up, once again, against the resistance of the Freudian concept of "psychic reality," which is the place where events are welcomed, transformed, translated, and even constituted. I would now like to undertake a lengthy study of this resistance in order to test—and thus to reinforce and to renew—the validity of claims that cerebrality has an etiological function.

The Crystal and the Psyche

In *New Introductory Lectures on Psychoanalysis*, Freud presents psychic disturbances as true ruptures that introduce real discontinuity into the life of
the subject. But the emergence of these ruptures always occurs along preexisting fault lines. *The psyche behaves exactly like a crystal*: it never shatters
haphazardly, but, within the disorder of its very shattering, always follows a
secret fault line that prepares it to shatter in a specific manner.

> If we throw a crystal to the floor, it breaks; but not into haphazard pieces. It
> comes apart along its lines of cleavage into fragments whose boundaries,
> though they were invisible, were predetermined by the crystal's structure
> [*durch die Struktur des Kristalls vorher bestimmt war*]. Mental patients are split
> and broken structures of this kind [*Solche rissige und gesprungene Strukturen
> sind auch Geisteskranken*].[1]

Freud thus discerns within illness a predictable relation between the hazard
of shock and the necessary form of breakage, a form that obeys lines of
fracture already inscribed within internal psychic life. The metaphor of the
crystal thus makes it possible to understand that, for Freud, the event is
always a synthesis between the unexpected accident and the endogenous
processes that produce psychic events.

It is precisely the originary possibility of such a synthesis that makes it
possible to uphold sexuality as a privileged etiological principle. To underscore the point once again, the concept of "sexuality," for Freud, names a
specific regime of events. "Sexual life" designates both a natural development of the psyche and the body and a set of events that come from outside
to impinge upon this unity of development. The events of sexuality solder
together chance and necessity and, for this reason, become the very model,
or the archetype, of every event. Sexuality derives its privileged causal character from a specific articulation—which constitutes its essence—of the
relation between the two dimensions of "predisposition (*Anlage*)" and contingency. Psychic damage or disturbance can only happen to a psyche that,
in a certain sense, is waiting for them.

There are events that, because they have no link to the intimacy of the
psyche, can cause radical transformations of the psyche that do not actualize
its preexisting potentialities. But Freud's work does not recognize the

destructive power of such events. Although Freud's conception of the event was constantly evolving—from *Studies on Hysteria* or "Sexuality in the Etiology of the Neuroses" to late texts such as *Beyond the Pleasure Principle*, *The Ego and the Id*, or *Outline of Psychoanalysis*—the one thing he never questioned was its character as a *synthesis*. From the theory of infantile trauma to the analysis of "disasters . . . involving a risk to life,"[2] Freud always upholds the double articulation of the event.

The wound by itself—the brain wound in particular—is unable to determine the event and is always subordinated to the *meaning* of the effraction. Compared to an electrical system in *Studies on Hysteria*, deprived of any role in the formation of psychic disturbances in "Introduction to *Psychoanalysis and the War Neuroses*," the brain is never acknowledged to possess etiological autonomy. It cannot, under its own power, articulate what comes from outside in relation to the course of endogenous events. There is thus no way to attribute any power of psychic creation to the brain's events because they are not, in reality, its own.

Freud does not minimize the impact of accidents, catastrophes, or disasters upon the ordinary course of the psyche's life. Far from it. From the theory of infantile sexual aggression to his reflections on war, he never ceases elaborating the notion of trauma: "We apply [the term 'traumatic'] to an experience that within a short period of time presents the mind with an increase in stimulus too powerful to be dealt with or worked off in the normal way, and thus must result in permanent disturbances of the way in which this energy operates."[3] Or again: "We describe as 'traumatic' any excitations from outside that are powerful enough to break through the protective shield (*Reizschutz*)."[4] However, in the same manner that Freud is led, very early on, to diminish the importance of "accidental factors" in the constitution of the neuroses, he just as rapidly minimizes the role of trauma itself even within the genesis of what he calls "traumatic" neuroses. Trauma is traumatic only to the extent that it triggers an internal conflict that exists before it. The enemy against which the soldier defends himself is always "an internal enemy."[5]

In order now to develop the Freudian conception of the event, my task is precisely to explore both the distance and proximity between the analysis of sexual accidents that occur in childhood and that of traumatic neurosis—in particular, war neurosis.

Traumatic Neurosis in Question

ACCIDENTS AND INTERNAL CONFLICTS

In *Beyond the Pleasure Principle*, Freud declares, "A condition has long been known and described which occurs after severe mechanical concussions, railway disasters and other accidents involving a risk to life; it has been given the name 'traumatic neurosis.'"[6] It must be recalled that, during the latter half of the nineteenth century, there were, in fact, many railway disasters and that the rail companies were held responsible for the consequences of these accidents. The concussed patients suffered a strange symptom that doctors called "railway spine." According to the reigning medical theories, the symptom was caused by a shock to the spinal cord, by a meningomyelitis of traumatic origin.

Hermann Oppenheim introduced the term "traumatic neurosis" in order to characterize the psychic consequences of such accidents.[7] It was not until a few years later that "war neurosis" was identified as an etiological variety of traumatic neurosis.[8] In response to this redefinition of neurosis, Freud also began to consider the psychic impact of accidental physical wounds and to take into account types of trauma other than childhood trauma directly attached to "sexuality" in the usual sense of the word. It was necessary for psychoanalytic theory to confront states of "fright [*Erschütterung*] . . . that a person gets into when he has run into danger without being prepared for it,"[9] even if the existence of such states seemed to call into question the supremacy of the pleasure principle.

Nonetheless, even when he takes into account these new forms of agitation and trauma, Freud does not call into question the sexual etiology of psychic disturbances, which remained unchanged since the beginnings of his work. Indeed, "mechanical agitation must be recognized as one of the sources of sexual excitation"[10] (see the remarks on the effects of rocking and train travel in *Three Essays on the Theory of Sexuality*). In the case of accidental mechanical shock, sexual excitation occurs without warning, since the psychic apparatus had no chance to steel itself against the assault; and, in reality, it is this unforeseen attack, not the shock itself, with its cerebral repercussions, that would generate anxiety and result in trauma as such. In other words, there is trauma only when the agitation produces nonmasterable sexual excitation that, by its violence, awakens a preexisting conflict.

In war neurosis, once again, trauma is not caused by shock but by the "ego conflict" that such shock triggers. "Shell shock," as war neurosis is often called in English, thus harks back to an older war than the one waged on the front.

FREUD ON THE FRONTLINES OF WAR NEUROSIS

Freud develops this idea in "Introduction to *Psychoanalysis and the War Neuroses.*" The year is 1915. Freud's three sons, his son-in-law, and his nephew have gone to the front. During this dark period, which will give birth to "Thoughts for the Times on War and Death," Freud thinks about writing *Beyond the Pleasure Principle* and is actively planning the Fifth International Psychoanalytic Congress, which would take place in Budapest on September 28 and 29, 1918, and whose theme would be precisely "psychoanalysis and the war neuroses."[11]

During this congress, Freud had to take a position about the nature of the traumatic illnesses, which were on the rise during the war, and about the possibility of their treatment with psychoanalysis. He begins by quite clearly supporting the idea that psychoanalysis would be the only effective treatment for such neuroses. Neurology, he affirms, is incapable of treating these disturbances to the extent that it does not take into account the fact that wounds, "patent" or not, and regardless of their organic character, always result from a sexual etiology. Accordingly, lesions to the nervous system, arising from certain head wounds, cannot by themselves provoke neurotic disturbances. It is, Freud writes, the *narcissistic withdrawal* of the libido, arising in the aftermath of the wound, the fact that "the sick person withdraws his libidinal investments into his ego," that causes the disturbance. As Louis Crocq remarks, "just as transference neuroses derive from the avatars of 'object libido,' which becomes fixated . . . upon an external love object, the war neuroses and the traumatic neuroses of peacetime derive from avatars of 'narcissistic libido' that attaches itself to the ego."[12]

Freud goes along with Ferenczi, who declared, during the Budapest Congress, that "the experiences among war neurotics gradually led further than the discovery of the mind: they led neurologists very nearly to the discovery of psychoanalysis." And also: "An advance, although one that is not admitted, has taken place in the attitude of leading neurologists toward psychoanalysis."[13] One thing is certain: no organic lesion can cause a war neurosis. The etiology of such neuroses is sexual which implies the conjunction of both sides—endogenous and exogenous—of the accident.

FREUD'S REPORT AS AN EXPERT WITNESS AT THE WAGNER-JAUREGG TRIAL

The position of psychoanalysis on the war neuroses received its final elaboration during an important event: Freud's participation as an expert witness called by the government in the trial of the military psychiatrist Julius Wagner von Jauregg, his colleague and friend. The proceedings of this trial are reported and analyzed in Kurt Eissler's remarkable book *Freud as an Expert Witness: The Discussion of the War Neuroses Between Freud and Wagner-Jauregg*.[14] The situation of the trial was as follows:

> In 1920, an exploratory committee organized a trial in Vienna in response to the complaint of a certain Lieutenant Kauders, who was gravely wounded and accused of malingering, in order to determine whether the military doctors during the war of 1914–18 were negligent in their use of electro-shock therapy—or "faradization"—on soldiers afflicted with psychic disturbances who had been identified as malingerers.[15]

Freud was thus summoned to bear witness as an expert before this committee. His testimony was recorded under the title "Memorandum on the Electrical Treatment of War Neurotics."[16]

Freud was upset during the trial because he did not want to malign a friend and an eminent personage within the medical world. (Not long after, Wagner-Jauregg would be totally rehabilitated).[17] At the same time, he was intent upon underscoring the inhuman but primarily inefficient character of electroshock therapy. Affirming that war neuroses have psychic and not directly functional origins, Freud demonstrates that electricity is of no help and that only psychoanalytic treatment can do anything to cure the disturbance. As Erik Porge writes in his introduction to the French edition of Eissler's book:

> The principal reproach that Freud addresses to Wagner-Jauregg is that he "defines malingering too broadly" and that he fails to recognize that most cases of so-called malingering are in fact cases of neurosis. Freud goes so far as to invert the claim that malingerers are not ill, saying instead that: "all neurotics are malingerers; they simulate without knowing it and that is their illness."[18]

It is incontestably a step forward to recognize the affective origin of war neuroses and to analyze them in terms of libidinal investments. These developments make it possible to delocalize the conflict, to avoid making the body into its only theater and the one agent of its cure. They make it possible, therefore, to avoid the neurological or neuropsychiatric reduction of psychic disturbances that prescribes shock treatment across the board—both in cases of malingering and of anxiety recognized as "real"—as the appropriate therapy for the nervous system.

SEXUAL ETIOLOGY AND THE "LIBIDO THEORY"

How would it be possible, then, in view of these arguments to establish the validity of cerebrality in the theory of trauma and of psychic events in general? Such an undertaking is far from straightforward. As we shall see, what Freud called the "sexual etiology of the neuroses"—of the traumatic neuroses in particular—is a highly differentiated and complex structure that is difficult to negotiate or to criticize without becoming entangled within it.

It does not suffice to recall that when Freud speaks of the "sexual" etiology of the neuroses, he means "sexuality" in an expanded sense. We must also explain this sense of the term.

Freud repeatedly indicates that his theory of psychic energy is *dualist*. It is precisely this divided energetics that he baptizes "libido theory." The "libido theory" always supposes difference among sources of energy. The libido is not the only energy present within the psyche. As his theory of drives evolved, Freud first divided the sources of energy between the sexual drive and the ego drive; and then, within the family of sexual drive, between the object drive and the ego drive (narcissism); and finally, in the final version of his theory, between the life drive (which comprises all the sexual drives) and the death drive.

In order to affirm that cerebrality resists all these differential turns of the "libido theory," therefore, it would be necessary to show that the psychic consequences of cerebral accidents and posttraumatic behavior in general cannot be entirely ascribed to one or an other of the different etiological principles contained within the same causal character. It would be necessary to make sure that the metamorphosis of identity without precedent is not the result of a super-subtle form of seduction, nor of a narcissistic fold, nor of a drive of aggressivity, nor of the internal work of annihilation immanent to the death drive understood as the return to the inanimate.

There is no way to get around all the barriers and pitfalls that the Freudian concept of sexuality opposes to the recognition of cerebrality. It is, therefore, only by situating myself at the very heart of Freud's thinking of the event, the accident, and trauma, attempting to do justice to the complexity of sexual etiology and the theory of the libido, that I deign to elaborate, under the sign of cerebrality, *a thinking of the destruction of the psyche different from that of psychoanalysis.*

To this end, it will be necessary to return to the frontlines of the war neuroses and to interrogate, once again, the case of Lieutenant Kauders, whose true pathology was not recognized either by the military doctors or by Freud despite the fact that it was very precisely diagnosed and recorded by the neurologists. Wounded on the battlefield, Kauders "was examined thoroughly and a head x-ray was taken. He was told that there was a partial fracture in the upper section of his skull and that a hematoma might also have formed." Why didn't Freud—who, on this point, behaved exactly like Wagner-Jauregg—say anything about this lesion? Why didn't he admit the psychically determining character of this cerebral trauma upon the behavior and attitudes of the wounded man, *as all indicators suggested that he do?* Wasn't it possible for psychoanalysis to accept and to begin its work with the neurological diagnosis rather than avoiding it and diminishing its importance? What exactly does psychoanalysis repress when it neutralizes neurology in this manner?

FIVE

What Is a Psychic Event?

The true etiology of the psychoneuroses does not lie in such precipitating causes.

— SIGMUND FREUD, "Sexuality in the Etiology of the Neuroses"

What is a *psychic event?* This question appears, in all its radicality, in Freud's first texts. The *Studies on Hysteria* show that the main problem posed by this pathology concerns its etiology—that is, the real nature of its *agents provocateurs.*[1] How much agency should one ascribe to the accident within the precipitation of hysteria and, more generally, within the neuroses? On the other hand, how much agency should one ascribe to *endogenous factors*? How does the "pathogenic effect" of an experience arise from the conjunction of an aleatory event and an internal danger?[2]

Freud's response to these fundamental questions is not univocal, and, in the years following his studies on hysteria, a series of transformations will displace his definition of the event. But these modifications will take place under the sign of sexuality, which, in the process, will gain its definitive stature and legitimacy as a principle and a causal character.

Psychoanalysis and Superstition: Chance, Sign, and Symptom

Freud's 1901 work *The Psychopathology of Everyday Life* furnishes precious insight into the way in which the psychoanalyst conceives the structure of psychic events. The chapters entitled "Symptomatic and Chance Actions" and "Determinism, Belief in Chance, and Superstition—Some Points of View" are particularly eloquent in this regard.

Freud recalls that the material upon which the psychoanalyst begins his work is constituted by bungled actions, slips of the tongue, and dreams— phenomena that are usually considered to be accidental, contingent (*zufäl- lig*), or the pure result of chance (*Zufall*), and thus cannot be anticipated or explained.

> The "chance" actions which are now to be discussed . . . appear on their own account, and are permitted because they are not suspected of having any aim or intention. We perform them "without thinking there is anything in them," "quite accidentally [*rein zufällig*]," "just to have something to do"; and such information, it is expected, will put an end to any inquiry into the significance of the action. In order to be able to enjoy this exceptional position, these actions, which no longer put forward the excuse of clumsiness, have to fulfill certain conditions: they must be *unobtrusive* and their effects must be slight [*sie müssen unauffälig und ihre Effekte müssen geringfügig sein*].[3]

As Freud underscores, these acts derive their exceptional position from their banality.

How does one understand the paradox that reveals the exceptional at the heart of the insignificant? In reality, these acts are *symptoms*. "I have col- lected a large number of such chance actions from myself and others, and after closely examining the different examples I have come to the conclu- sion that the name of *symptomatic acts* is a better one for them [*Symptomhandlungen*]."[4] The word "symptom" is particularly clarifying because it designates a phenomenon that is both a matter of *chance* and *necessity*. In Greek, *symptoma*, refers first of all to the wilting or collapse of something that falls (something that actually happens and *falls* upon us), the coincidence, the fortuitous incident, and finally to the sign that both with- holds and reveals its meaning. The symptom is situated at the intersection of the accident and hermeneutics: it is a *meaningful accident*. It constitutes an accident that is tantamount to a confession: forgetting things, putting stains on clothing, dropping things on the floor, misremembering things, and so on. Freud thus confronts an urgent question: Does not conferring symptomatic value upon such incidents *make one superstitious*? Is not psychoanalysis purely and simply a form of superstition that endows merely fortuitous events with the meaning and value of coincidences?

Chapter 12 develops one of the most striking examples of "symptomatic acts." Freud tells us that, upon returning from vacation, he had to go by carriage to visit one of his very aged patients:

> She is over ninety years old; it is therefore natural to ask oneself at the beginning of each year's treatment how much longer she is likely to live. On the day I am speaking about I was in a hurry and called a cab to take me to her house. Every cabman on the rank in front of my house knew the old lady's address, as they had often taken me there. But on this day it happened that the cabman did not draw up in front of her house but in front of a house with the same number in a nearby street which ran parallel and was in fact of similar appearance. I noticed the error and reproached the cabman with it and he apologized.[5]

Can this error about the address be considered mere chance or does it mean something? Freud responds:

> Now is it any significance that I was driven to a house where the old lady was not to be found? Certainly not to me, but if I were superstitious I should see an omen in the incident, the finger of fate announcing that this year would be the old lady's last.[6]

The superstitious person and the psychoanalyst do indeed share a point in common: Neither one believes in chance, so they share the penchant for interpretation. Freud continues:

> I am therefore different from a superstitious person in the following way:
>   I do not believe that an event [*Ereignis*] in whose occurrence my mental life plays no part can teach me anything about the future shape of reality; but I believe that an unintentional manifestation of my own mental activity does on the other hand disclose something hidden [*Verborgen*], though again it is something that belongs only to my mental life. I believe in external (real) chance, it is true, but not in internal (psychical) accidental events [*ich glaube zwar an aüsseren (realen) Zufall, aber nicht an innere (psychische) Zufälligkeit)*].[7]

This remarkable conclusion makes it possible to grasp the fundamental distinction between the event such as it occurs from outside, in an entirely unanticipated fashion—for which Freud uses the word *Ereignis*—and the *lived* event, that which becomes meaningful for the psyche, for which Freud uses the word *Erlebnis*. The superstitious man makes every *Ereignis*—which, here, designates every accidental event—into a sign or symptom. The psychoanalyst, however, restricts himself exclusively to the domain of *Erlebnisse*—the lived events, meaningful for the psyche, or, if one prefers, the "subjective" events that constitute the symptomatic dimension. Accordingly, Freud would not have conferred significance upon incident (*Vorgang*), in itself insignificant, of the wrong address unless he had made the mistake *himself*, unless his own psyche, and not that of the cabman, had somehow ordained it.

The first definition of the psychic event thus lies in the conjunction of *Ereignis* and *Erlebnis*. Unless the one can be "translated" into the other, no event is marked as such—which is to say, as a symptom.

The Sexual as Transformer

How is this conjunction or this translation produced? How can the absence of internal chance coincide with the presence of "real" chance? The link between external accident and endogenous event, between inside and outside, chance and necessity, contingency and meaning, is what, strangely, in Freud's work will be called *sexuality*. Indeed, "sexuality" is the point of transition from one signification of the event to the other. Playing a bit with Lacan's famous formula, we could affirm that, if "there is no sexual relation," then the sexual is still, before all else, *the name for a relation*. Once again, sexuality does not designate sexual experiences of practices but the constitution of a certain type of eventality. More precisely, sexuality is at the origin of the conjunction between *Ereignis* and *Erlebnis*.

STUDIES ON HYSTERIA: THE ETIOLOGICAL FACTOR OF HYSTERIA
ACCORDING TO BREUER

This conjunction was perceived as early as *Studies on Hysteria*. Freud's contribution could be distinguished from Breuer's in that he explicitly took this conjunction into account. In the preface to the second edition of *Studies*, Freud notes the "developments and changes" of his "views" on the sexual etiology of hysteria.[8]

In the preface to the first edition of the *Studies*, Freud and Breuer wrote: "Sexuality seems to play a principal part [*Hauptrolle*] in the pathogenesis of hysteria as a source of psychical traumas [*als Quelle psychischer Traumen*] and as a motive [*Motiv*] for 'defense'—that is, for repressing ideas from consciousness."⁹ The role of sexuality was thus clearly perceived from the beginning, but it still did not have the same importance for Breuer as it had for Freud. In fact, Breuer's contribution remains almost entirely silent on the question of sexuality, except for one remark: "We are already recognizing sexuality as one of the major components of hysteria."¹⁰ In his contribution, however, Freud presents decisive elaborations—which mark a point of rupture between himself and Breuer—of the constitution of sexuality as a regime of events and, consequently, as a determining cause of hysteria and of neurosis in general.

Breuer begins by proposing, as he articulated it with Freud in the "Preliminary Communication," to establish an analogy between "ordinary hysteria" and "traumatic neurosis" in order to found the concept of "traumatic hysteria." In the case of traumatic neurosis, the neurosis is occasioned by a wound but also, more important, by the fright that accompanies it:

> In traumatic neuroses the operative cause is not the trifling physical injury but the affect of fright—the psychical trauma. In an analogous manner, our investigations reveal, for many, if not for most, hysterical symptoms, precipitating causes which can only be described as psychical traumas. Any experience which calls up to distressing affects—such as those of fright, anxiety, shame or psychical pain—may operate as a trauma of this kind.¹¹

In all cases of hysteria, therefore, there is the equivalent of a traumatic *shock*.

What is this equivalent? Herein lies the entire question. What, at bottom, is a *wound* of a psychic order? Strangely, Breuer is not very clear on this point. He insists on the "accidental factor [*die akzidentelle Moment*]" and the "precipitating trauma [*veranlassendes Trauma*]" within psychic disturbances but does not specify their nature. The only point that seems clearly established is the fact that the accidental factor has causal value: "It is, of course, obvious that in cases of 'traumatic' hysteria what provokes the symptoms is the accident."¹² In addition, "the symptoms disappeared immediately and did not recur if we succeeded in wakening the memory of the precipitating event [*veranlassende Vorgang*]"—an event that is also sometimes called the "precipitating cause [*auslösende Ursache*]."¹³

The "precipitating" event must nonetheless be linked to affect in order to have an effective traumatic impact. It must coincide with strong paralyzing emotions. Trauma thus proceeds from an encounter between shock and affect:

> For we found, to our great surprise at first, *that each individual hysterical symptom immediately and permanently disappeared when we had succeeded in bringing clearly to light the memory of the event by which they were provoked and in arousing their accompanying affect, and when the patient had described that event in the greatest possible detail and had put it into words*. Recollection without affect almost invariably produces no result.[14]

There is thus an encounter between the unexpected accident and endogenous excitation. This encounter forms the event that Breuer calls a "foreign body" that perturbs nervous equilibrium: "psychical trauma—or more precisely, the memory of the trauma—acts like a foreign body [*Fremdkörper*] which long after its entry must continue to be regarded as an agent that is still at work."[15]

In order to explain the coupling of external incident and affect as well as the constitution of the foreign body, Breuer shows that the incident provokes an excessive influx of excitation within the nervous system, an influx that it cannot manage to liquidate. Indeed, the hysterical shock produces what Breuer calls "intracerebral tonic excitation" and explicitly considers as an "inflammation of the brain."[16]

It is at this point in his argument that he develops the comparison between the nervous system and an electrical grid:

> Let us imagine a widely ramified electrical system for lighting and the transmission of motor power; what is expected of this system is that simple establishment of a contact shall be able to set any lamp or machine in operation. To make this possible, so that everything shall be ready to work, there must be a certain tension present throughout the entire network of lines of conduction, and the dynamo engine must expend a given quantity of energy for this purpose. In just the same way there is a certain amount of excitation present in the conductive paths of the brain when it is at rest but awake and prepared to work.[17]

Surpassing the optimum level of excitation results, as we have seen, in a "short circuit"[18] within the system. The psychical equivalent of a short-circuit is the creation of an "abnormal facilitation" due to the "the tension of cerebral excitation . . . [passing] over to the peripheral organs."[19] Breuer concludes: "a non-uniform distribution of increased excitation is what makes up the psychical side of the affects."[20]

This excessive influx thereby turns into a symptom, that is, into a somatic manifestation. This is the phenomenon of *conversion*. The coupling of incident and affect is due to the coincidence within the nervous system between (1) a shock received from the outside; (2) the emotional excess that it incites; and (3) the incapacity of the nervous system to discharge this excitation that finally turns into a symptom:

> The experiences which released the original affect, the excitation of which was then converted into somatic phenomenon, are described by us as psychical traumas, and the pathological manifestation arising in this way, as *hysterical symptoms of traumatic origin.*[21]

THE FREUDIAN INTERVENTION: THE FIRST DELOCALIZATION FROM THE
CEREBRAL TO THE SEXUAL AND THE FIRST VERSION OF "SOLDERING"

When Freud speaks in his own name, he accomplishes a remarkable radicalization of the etiological problem:

> Thus, starting out from Breuer's method, I found myself engaged in a consideration of the etiology and mechanism of the neuroses [*Ätiologie und Mechanismus der Neurosen*]. . . . In the first place, I was obliged to recognize that, in so far as one can speak of determining causes [*Verursachung*] which lead to the acquisition of neuroses, their etiology is to be looked for in sexual factors [*sexuellen Momenten*]. There followed the discovery that different sexual factors, in the most general sense, produce different pictures of neurotic disorders.[22]

Freud does not at all modify Breuer's comparison between the nervous system and an electrical grid. He even puts it to use. He merely denies that the cerebral "circuit" has any primacy within the process of managing events. He declares that it is necessary to found, outside the circuit, a topic of the accident—which amounts to definitively separating this topic, as an etiological structure, from the nervous system itself. The latter thereby becomes merely a means of transmitting or managing, but not forming, the synthesis between *Ereignis* and *Erlebnis*. The formation of this synthesis must now take place elsewhere.

The radicalization of sexuality as an etiological principle thus consists in a delocalization that results at once from a conservation and a suppression. Even as he retains the definition of the nervous system as a circuit, Freud also continues to uphold the idea—which he will never abandon—that an accident, regardless of its force, is incapable on its own of causing a perturbation of an energetic economy.[23] It is thus surprising that he will employ the lexicon of energetics to describe both the functioning of the *nervous system* and the vicissitudes of *sexual excitation*. To speak of sexuality, indeed, is to speak of excitation, tension, liquidation, and discharge or, on the contrary, of inhibition. In a certain sense, the metaphor of the electrical grid is perfectly suited to characterize sexual economy. It is thus a similarity of register—the relation between tension and discharge—that authorizes the substitution of the sexual for the cerebral.

But it is at the site of this similarity that the two will be forever separated. Displacing the causal authority of energetic perturbation, Freud introduces a symbolic dimension into the economy of the accident that is foreign to the nervous system but constitutive of the sexual functioning. This dimension concerns precisely the conjunction of *Ereignis* and *Erlebnis*—that is, the dual structure of the psychic event that, in "Hysterical Fantasies and Their Relation to Bisexuality," Freud calls "soldering [*Verlötung*]."[24] Between one type of event (the "incident") and another (the event that becomes significant for the psyche), Freud situates an element—a veritable transformer—that never has its source in the nervous system and remains radically foreign to it: the *fantasy* (*Phantasie*). Sexuality thus names a regime of eventality that is situated at the intersection of an *energetic circuit* and a *theatrical scene*—which implicitly denies that the nervous system can have the status of a stage. In "Some General Remarks on Hysterical Attacks," Freud declares that hysterical "attacks" (*Anfallen*—a word that, we will see, designates both the unforeseen upsurge of mental representations and warlike assaults) "are nothing else but fantasies translated into the motor sphere, projected onto motility and portrayed in pantomime."[25]

Without this fantasmatic soldering, there would be no internalization of shock or translation of it into affect; this place that Lacan called an "intimate exteriority," the proper location of psychic events, would never be constituted.[26] Sexuality thus becomes what cerebrality is incapable of being: an outside of the inside that is not foreign to this inside itself—an outside of the inside *proper* to the psyche.

If, in subsequent works, Freud never ceases to reevaluate the parts played by *Ereignis*, *Erlebnis*, and fantasy in the formation of psychic accidentality, he never calls into question the fact, posited in *Studies*, that *sexual accidents, and thus their power of "soldering," determines the structure of every psychic event.* Throughout Freud's work, the "determining force [*Eignung zur Determinierung*]" or the "determining power [*betreffende determienende Eignung*]"[27] of trauma is shown to reside in this soldering.

Freud will then explain the *pyramidal* character of the causal series linked to this "determining force":

> Whatever case and whatever symptom we take as our point of departure, *in the end we infallibly come to the field of sexual experience.* So here for the first time we seem to have discovered an etiological precondition.[28]

By the same token, the symptom can no longer be described as a foreign body. Instead of this metaphor, which Breuer used, Freud prefers that of the *infiltrate*:

> I shall now make one or two further remarks on the picture we have just arrived at of the organization of the pathogenic material. We have said that this material behaves like a foreign body, and that the treatment, too, works like the removal of a foreign body from the living tissue. We are now in a position to see where this comparison fails. A foreign body does not enter into any relation with the layers of tissue that surround it, although it modifies them and necessitates a reactive inflammation in them. Our pathogenic psychical group, on the other hand, does not admit of being cleanly extirpated from the ego. Its external strata pass over in every direction into portions of the normal ego; and, indeed, they belong to the latter just as much as to the pathogenic organization. . . . The interior layers of the pathogenic organization are increasingly alien to the ego, but once more without there being any visible boundary at which the pathogenic material begins. In fact the pathogenic organization does not behave like a foreign body, but far more like an infiltrate. . . . Nor does the treatment consist in extirpating something— psychotherapy is not able to do this for the present—but in causing the resistance to melt and thus enabling the circulation to make its way into a region that has hitherto been cut off.[29]

The soldering of external events and lived experience gives the event itself *historical* value: thanks to this process, it is integrated into the history of the subject and can never be isolated from it. After its transformation, the event becomes unlocalizable because infused with subjectivity. Its presence is no longer exotic: once a stranger, it has become the subject's own.

*The Theory of Seduction and the Sexual Traumas of Childhood:
Evolution and Withdrawal of the* Ereignis

In its turn, this first version of the conjunction between *Ereignis* and *Erlebnis*—of "soldering"—will be transformed. It is difficult not to observe that the relative importance of the two evental forces becomes unbalanced. *The power of Ereignis recedes in favor of Erlebnis.* This mutation is particularly obvious in Freud's transition from his first theory of infantile seduction to his second theory. In the latter, Freud tends to minimize the part played by *actual* sexual aggression in early childhood and their impact within the history of the subject. The notion of sexual accident thereby changes meaning and, along with it, the notion of the event.

FIRST THEORY: "ACTUAL" SEXUAL ABUSE

Let us begin with the theory of the seduction of the child. In "The Etiology of Hysteria," in 1896, Freud declares to his audience of physicians:

> As you know, in the view of the influential school of Charcot, heredity alone deserves to be recognized as the true cause of hysteria, while all other noxae of the most various nature and intensity only play the part of incidental causes [*Gelegenheitursachen*], of "*agents provocateurs.*"[30]

However, heredity cannot be the sole etiological factor. The proof is that certain neuropathies can develop in "people who are perfectly healthy and whose family is above reproach."[31]

Freud recalls that there are in fact three principal causes of neuroses: the "conditions" (heredity foremost among them), the "concurrent causes" (moral emotions, exhaustion, overwork, and traumatic accidents—not forgetting the secondary character of the latter), and last, "specific causes," which pertain to the individual's sexual life.[32] "Pathological modifications," Freud specifies, are linked to "*important events in* [the subject's] *past life.*"[33]

These "important events," which occur in childhood, are acts of sexual aggression. The sexual accident is here understood as *Ereignis*, as an *effraction* coming from outside. Certainly, fantasy plays the role of soldering agent but it does not reduce the role of external accidentality. "It seems to me certain," Freud writes, "that our children are far more often exposed to sexual assaults than the few precautions taken by parents in this connection would lead us to expect."[34] He then goes on to enumerate the three most common types of aggression: (1) attacks or abuse committed by "adults who were strangers to the family"; (2) abuse by people close to the family (maid, preceptor, or parent); and finally, (3) abuse perpetrated by other children and premature amorous relations of children among themselves.[35] The seriousness of these aggressions ranges from "astonishingly trivial" events to "severe traumas" such as rape or incest.[36] Freud concludes:

> It is not the latest slight—which, in itself, is minimal—that produces the fit of crying, the outburst of despair or the attempt at suicide, in disregard of the axiom that the effect must be proportionate to the cause; the small slight of the present moment has aroused and set working the memories of very many, more intense, earlier slights, behind all of which there lies in addition the memory of a serious slight in childhood which has never been overcome.[37]

Underlying every case of hysteria, one finds one or many events (*Ereignisse*) that pertain to "premature sexual experience [*vorzeitiger Erfahrung*]" dating from the earliest years of childhood. "I believe that this is an important finding, the discovery of a *caput Nili* [source of the Nile] in neuropathology."[38] Sexuality, as we can see, is the synthesis of the actual event and its internalization.

SECOND VERSION OF THE CONJUNCTION BETWEEN *EREIGNIS* AND *ERLEBNIS*: "PSYCHIC REALITY"

We should recall that, from the *Studies on Hysteria* onward, Freud declares that "hysterics suffer from reminiscences" and that, if the event of an effraction can rightly be called traumatic, the memory of the effraction, its "aftershock," is equally determining. The mnesic elaboration of the traumatizing scene thus appears just as important, in its value as a source, as the occurrence of the scene itself. At the same time, the first scene only becomes pathogenic within memory, which, unto itself, produces an a posteriori influx of excitation. Throughout his work, Freud thus will be led to nuance his account of the role played by *Ereignis*, or the external event, within the formation of trauma. The "soldering" does not break apart but is formed otherwise.

Freud will ultimately defend the idea that external events derive their efficaciousness from the fantasies that they activate and from the excitation that these fantasies trigger. Accordingly, fantasy no longer appears as the hinge or point of conjunction between *Ereignis* and *Erlebnis*, between what arises from outside and the lived event. On the contrary, fantasy constitutes their indissoluble, even identical, character, their originary unity, which thus no longer requires soldering. Fantasy itself (*Phantasie*) comes to designate the mode in which the accident occurs within the psyche. It is no longer a matter of the incident on one side and its reception within the intimacy of the psyche on the other. The psychic event, from then on, is linked to *an originarily fantasmatic occurrence of the accident.*

Freud affirms that it is fantasy that, in a sense, makes come what comes—makes it come back at the very moment that it surges up—such that the event becomes an incisive point that happens and inscribes itself at the same time. "Soldering" then changes meaning. Whereas it began as the suture between *Ereignis* and *Erlebnis*, it becomes the conjunction of two differentiated instances at the heart of the same apparatus which Freud calls "psychic reality [*psychische Realität*]" and which he is careful to distinguish from "material reality." In "On the History of the Psychoanalytic Movement," he writes:

> If hysterical subjects trace their symptoms to traumas that are fictitious, then the new fact which emerges is precisely that they create such scenes in fantasy, and this psychical reality requires to be taken into account alongside practical reality.[39]

This affirmation echoes another from *The Interpretation of Dreams* according to which "psychical reality is a particular form of existence not to be confused with material reality."[40] Freud begins by admitting the material or factual reality of childhood scenes but he subsequently abandons this initial conviction, showing that such reality is in fact a "psychical" reality. The traumatic event is not "a real fact" in the sense of "material reality,"[41] but this does not mean that it is a lesser reality. It is "other."

What does this mean? There are "two" realities but "psychical reality" is itself twofold. The distinction between the unexpected occurrence of the event and the reception of this event within the intimacy of the psyche now takes place within psychical reality itself. Fantasy becomes the conjunction between an internal upsurge—the accident occurs from inside with as much force as from outside—and its interpretation or its repetition within an internal fiction elaborated by the individual psyche. The distinction between inside and outside no longer corresponds exclusively to the division between psychical reality and material reality but to the internal division of psychical reality itself. "Soldering" thus becomes inseparable from the structure of this reality.

The distinction between "psychical reality" and "material reality" solidifies the derivative or secondary status of the "real" accident (materially real) with respect to the "fantasmatic" accident (real otherwise) and firmly entrenches the subordination of "triggering factor" to "precipitating cause." An incident that occurs in "material reality" can trigger a disturbance without at all being its cause. One accident can hide another. An event—*Ereignis*—coming from "material" reality can provoke psychic damage without being the ultimate reason for this damage. It "triggers" another source of trouble, which is truly primary, inscribed within the psyche as a fantasmatic event. This duality of triggering and precipitating will remain a pregnant resource throughout Freud's work.

BETWEEN PSYCHICAL REALITY AND MATERIAL REALITY:
THE NON-PLACE OF CEREBRAL REALITY

How is it possible to defend the idea of a *fantasmatic* effectiveness of events? It is necessary to distinguish between two levels within Freud's understanding of *Phantasie*. The first level concerns "daydreams" which, in *Studies on Hysteria*, are defined as hypnoid states which include all the stories and fictions that the subject tells herself when she takes leave of "material reality." In *The Interpretation of Dreams*, Freud first classes fantasy among "diurnal phenomena," the structures of which are supposedly analogous to the nocturnal phenomena of dreams. The second level concerns fantasy as the *core of the dream*, the oldest part of the dream formation, the "navel" that Freud evokes in Chapter 7 of the work. Fantasy, in this sense, comes to designate the unconscious evental scenario, the upsurge of an internal accident inseparable from its pantomimic figuration. The dream, as Freud affirms especially in "Some General Remarks on Hysterical Attacks" (1909), "takes the place of an [hysterical] attack." The soldering at work in "psychical reality" thus consists in the conjunction between the fantasmatic accident conceived as an attack and its "hallucinatory distortions."[42]

For Freud, the nervous system is foreign to this fantasmatic staging that does perfectly well without it—does not collaborate in it, is not its topic, nor its author. At the moment when he takes up the question of the dreamwork, Freud shows precisely that this work is independent of cerebral processes. *There is a spontaneous psychic evental creation that is not produced by the operation of the nervous system.* "Modern psychiatry" has no idea about such creation since it views the brain as the only source of the dream and thus fails to recognize the "causal link"—the true etiological principle—that constitutes psychic life:

> We shall find later that the enigma of the formation of dreams can be solved by the revelation of an unsuspected psychical source of stimulation. Meanwhile we shall feel no surprise at the over-estimation of the part played in forming dreams by stimuli which do not arise from mental life. Not only are they easy to discover and even open to experimental confirmation; but the somatic view of the origin of dreams is completely in line with the prevailing trend of thought in psychiatry today. It is true that the dominance of the brain over the organism is asserted with apparent confidence. Nevertheless, anything that might indicate that mental life is in any way independent of demonstrable organic changes or that its manifestations are in any way spontaneous alarms the modern psychiatrist, as though a recognition of such things would inevitably bring back the days of the Philosophy of Nature, and of the metaphysical view of the nature of the mind. The suspicions of psychiatrists have put the mind, as it were, under tutelage, and they now insist that none of its impulses shall be allowed to suggest that it has any means of its own. This behavior only shows how little trust they really have in the validity of a causal connection between the somatic and the mental.[43]

The psychical regime of events, for Freud, is autonomous; it does not depend upon any organic causes—especially not upon any cerebral cause. This autonomy manifests itself precisely through the independence of fantasmatic work whose only creative resources come from the psyche and not the brain. Once again, the concepts of scene, fiction, and scenario are foreign to any neuronal organization that, according to Freud, does not possess an apparatus of representation.

This point already appears quite clearly in the text entitled "Some Points for a Comparative Study of Organic and Hysterical Motor Paralyses" (1893) where Freud finds that hysterical attacks are endowed with the capacity to *imitate brain damage*: "Hysteria has fairly often been credited with a faculty for simulating the most various organic nervous disorders"—in particular certain forms of paralysis.[44] This mimetic faculty bears witness precisely to the autonomy of psychic processes with respect to the cerebral system.[45] Hysteria, by virtue of its capacity for fiction, reconstitutes the body and its reactions according to a rule that entirely disregards anatomy: "I . . . assert that the lesion in hysterical paralyses must be completely independent of the anatomy of the nervous system, since in its paralyses and other manifestations hysteria behaves as though anatomy did not exist or as though it had no knowledge of it."[46] Hysteria mimes cerebral paralysis even as it caricatures it, both exaggerating and dissociating its symptoms. Such paralysis is characterized by its "excessive intensity" and at the same time by the "precise limitation" of its symptoms to isolated parts of the body, while cerebral paralysis would have affected them together.[47] The very particular redistribution of the illness is explained by the fact that hysteria refigures or reconfigures the whole of the body by attributing different affective values to each of its parts. The more the organ is invested, the more it becomes "heavy" or "inert."[48] Affect thus inscribes a body upon the body; and the second body results from an eroticization of the first, a phenomenon that, once again, only sexuality can account for.

We thus arrive at a decisive point. The evolution of his ideas that Freud would describe in "My Views on the Part Played by Sexuality in the Etiology of the Neuroses" (1905) was already secretly underway in his earliest texts. He admits that, early on, in his first theory of seduction, he "over-estimated the frequency of such events (though in other respects they were not open to doubt). Moreover, I was at that period unable to distinguish with certainty between falsifications made by hysterics in their memories of childhood and traces of real events [*der Spuren der wirklichen Vorgänge*]. Since then I have learned to explain a number of fantasies of seduction as attempts at fending off memories of the subject's own sexual activity (infantile masturbation)." From that point onward, he continues, "the 'traumatic' element in the sexual experiences of childhood lost its importance."[49]

With this shift of emphasis, Freud "ceased to lay exaggerated stress upon the accidental influencing of sexuality [*akzidentellen Beeinflüssung der Sexualität*]." It thus becomes clear that the neuroses are caused exclusively by "noxae that affect the sexual function itself" and that "noxae and trauma which [cause] general damage to the organism may lead [only] secondarily to injury to its sexual processes."[50]

The reelaboration of the theory of seduction, linked to the evolution of the concept of the psychic event, would then reveal its true scope. The avowed or unavowed goal of explicating the two types of reality—practical or material, on the one hand, and psychical, on the other—is to deny any specific reality to cerebral processes, to deprive cerebrality of its own regime of events.

Certainly, Freud insists on the fact that the "diminishing importance of accidental influences" within his theory of the psychic event is not tantamount to the pure and simple elimination of these influences. The incalculable accident, unforeseen catastrophe, and wounds occurring in "material reality" obviously remain important factors in psychic life. However, in the course of Freud's theoretical reelaborations, it is the notion of the cerebral event that gets stripped of any effectivity of its own. Indeed, insofar as every *Ereignis* is joined with an *Erlebnis* of equal importance, brain lesions could be invested with meaning and acquire a psychic dimension. With the barrier of fantasy, this equivalence is shattered. The psychic fissure that exists before any damage or any tumult now runs through fantasy itself. Sexual etiology is thereby definitively imposed within psychoanalytic theory.[51]

From the etiology of hysteria to that of the neuroses, sexuality is invested with the authority of law and causal character. In *An Outline of Psychoanalysis*, Freud shows that if, in theory, "any sort of libidinal demand might occasion the same repressions" and the same consequences as those caused by the sexual drive, "our observation shows us invariably, so far as we can judge, that the excitations that play this pathogenic part arise from the partial drives of sexual life."[52] The neuroses, therefore, must all be traced to their origins that concern the "noxae" impinging upon the sexual function.

It is clear that the causal position of sexuality, in Freud, is solidified at the price of a constant critique of the etiological claims of neurology and neuropathology. Between neurosis and brain damage, there can be points of contact, but never, precisely, a relation of cause and effect. Organic lesions, no matter what they are, cannot create a neurosis—that is, they cannot join an event and a fiction.

These considerations bring us to the fundamental question that Freud addresses retroactively to contemporary neurology. The source of narrativity, which is necessary for the presentation of pathological cases, lies in the fictioning power of illness itself. But if illness is stripped of such power, is it still possible, without the mediation of a fantasmatic work, to make it speak?

SIX

The "Libido Theory" and the Otherness of the Sexual to Itself:
Traumatic Neurosis and War Neurosis in Question

The psychoanalytic view of the traumatic neurosis [is not] identical with shock theory in its crudest form. The latter regards the essence of shock as being the direct damage to the molecular structure or even the histological structure of the elements of the nervous system; whereas what we seek to understand are the effects produced on the organ of the mind by the breach in the shield against stimuli and by the problems that follow in its train.

— SIGMUND FREUD, *Beyond the Pleasure Principle*

The concept of narcissism made it possible to obtain an analytic understanding of the traumatic neuroses and of many of the affections bordering on the psychoses, as well as of the latter themselves.

— SIGMUND FREUD, *Civilization and Its Discontents*

It is not easy to have done with the etiological primacy of sexuality. As the preceding analyses have shown, and as Freud indicates through his many warnings not to reduce psychoanalysis to "pansexualism," establishing the causal character of sexuality requires a complex elaboration.[1]

Even though sexuality is nomothetic within its order—it posits the law of its own field—*it remains differentiated in its structure*. The concept of sexuality functions to highlight the series of divisions and articulations of which it is constituted. For the moment, I have only taken into account the dual character of the psychic event, its "soldering," whose organization, in spite of the theoretical displacements that it undergoes, is always constituted as a diptych or a fracture: on one hand, the conjunction of *Ereignis* and *Erlebnis* and, on the other, the fantasmatic accident and its interpretive reception.

But another dualism, equally fundamental, underlies the binary constitution of sexuality: what Freud calls the *dualism of the drive*—in other words, the "libido theory." It is this dualism that reveals *the full complexity of sexual etiology* and that underscores yet again the difficulties that stand in the way of according cerebrality its own autonomy and its own evental power. An exhaustive examination of these difficulties requires that I elaborate the theory of the dualism of the drive from the perspective of the Freudian conception of trauma, showing once again that the armature of "sexual" eventality, which Freud reinforces from text to text, can apparently be shaken by the destructive occurrence of an unforeseen catastrophe.

Sexuality as the Otherness of the Sexual to Itself

THE DUALISM OF THE DRIVE AGAINST THE MONISM OF ENERGY

Contrary to what one might tend to think, sexuality, for Freud, never functions as a *monological* cause but designates a difference, a couple, a bifid conceptual structure. In this sense, however one decides to approach it, *sexuality always reveals the otherness of the sexual to itself*; and the theory of the dualism of the drive, which is inseparable from what Freud calls the "libido theory," manifestly validates this observation.

Throughout his work, Freud presents the theory of the drive as a dualism. If every neurosis, in the last instance, can be derived from sexual causes, this does not mean that the sexual drive is not internally differentiated. Within the drive, the sexual and the other of the sexual are always at work together. The "sexual" is always one half of a couple.

What does this mean? Throughout his work, Freud theorizes the drive as a dualism. In "Drives and Their Vicissitudes," Freud presents the first version of his libido theory, which divides the drive into "ego drives" and "sexual drives." The former impel the individual to preserve himself; the latter, on the contrary, push him to go beyond his own limits and to reproduce himself: "On the one hand, the individual is the principal thing . . . while, on the other, the individual is a temporary and transient appendage to the immortal germ-plasm, which is entrusted to him by the process of generation." Accordingly, sexual aims, contrary to the aims of the ego, "go beyond the individual and have as their function the production of new individuals—that is, the preservation of the species."[2]

In 1917, in "A Difficulty in the Path of Psychoanalysis," Freud comments on this distinction:

> [I]n psychoanalysis . . . we make a distinction between the self-preservative or ego drives on the one hand and the sexual drives on the other. The force by which the sexual drive is represented in the mind we call "libido"—sexual desire—and we regard it as something analogous to hunger, the will to power, and so on, where the ego drives are concerned.[3]

Later, Freud will remark that the ego itself, far from being the "other" of the sexual drives, can also become an object of libidinal investment. In the second version of the dualism of the drive, therefore, he establishes and elaborates the libidinal content of narcissism. Accordingly, the new theory will no longer rest upon the distinction between ego drives and sexual drives, but rather upon the distinction between "ego drives" and "object drives." Commenting on this evolution, the author writes:

> Advancing more cautiously, psychoanalysis observed the regularity with which libido is withdrawn from the object and directed on to the ego (the process of introversion). . . . The ego now found its position among sexual objects and was at once given the foremost place among them. Libido which was in this way lodged in the ego was described as "narcissistic."[4]

Finally, in 1920, in *Beyond the Pleasure Principle*, offering a final elaboration of the dualism, Freud unites ego drives and sexual drives within the same group—that of sexual drives (also called "life drives")—and defines the other of the sexual as the "death drive."

DIFFERENCE IN THE LIBIDO

The dualism of the drive is directly linked to a theoretical division, essential in Freud, of the concept of the libido. There are a "weak" sense and a "strong" sense of the concept of libido. In the "weak" sense, the libido designates the "dynamic manifestation of the sexual drive." In the "strong" sense, the libido designates less a specific energy than a global theory of energy, according to which, precisely and paradoxically, the energy of the drive is not merely libidinal. "Something in the nature of a theory has at last shaped itself in psychoanalysis, and this is known by the name of the 'libido theory,'"[5] Freud writes. And this "theory" corresponds to the dualism of the drive.

In spite of what the term might lead one to conclude, the "libido theory" must not be understood as a "theory of sexual desire or pleasure" but rather as a theory of *the partition of energy*. Even if, as we saw at the beginning, the libido functions as an archetype that unites all psychic energy under the same sign, this unification is only of heuristic value.[6] In reality, the libido is only one of the energies at work within the psyche. To speak of the "libido theory" is tantamount to admitting that the libido is not the only thing at stake. The libido theory is paradoxically a theory of the non-self-sufficiency of the libido. The sexual is always bound up with its other. This otherness, however, does not diminish but rather consolidates the etiological power of sexuality.

Sexuality, as a causal principle, always has more than one origin. Accordingly, to "subsume" a case or disturbance "under the hypothesis of the libido theory [*durch die Libidotheorie ermöglichen*]"[7] is to decide which element of the couple, libidinal or other, determined its genesis and its development. To explain a disorder in terms of sexual etiology is never a matter of abusively "reducing" it to the sexual but, on the contrary, of sifting it through a difference, the difference of the sexual to itself—whether this difference is called the ego or the death drive.

On many occasions, Freud reasserts his total disagreement with Jung on this point. Jung, in his view, had "enlarged" the concept of the libido to make it into the single energy at work within the psyche—an energy that could, if necessary, be "sexualized" or "desexualized."[8] Freud declares:

> Our views have from the very first been *dualistic*, and today they are even more definitively dualistic than before—now that we describe the opposition as being, not between ego drives and sexual drives but between life drives and death drives. Jung's libido theory is on the contrary monistic; the fact that he has called the singular force of the drive "libido" is bound to cause confusion, but need not affect us otherwise.[9]

In his encyclopedia article on the libido theory from 1923, Freud further sharpens his critique of Jung: "C. G. Jung attempted to resolve this obscurity along speculative lines by assuming that there was only a single primal libido which could be either sexualized or desexualized and which therefore coincided in its essence with mental energy in general." This hypothesis is inadmissible, for Freud, because it tends to abolish the structural difference that constitutes the drive, while it remains incontestable that "the difference between the sexual drives and drives with other aims was not got rid of by means of a new definition."[10]

Examining all the versions of the dualism of the drive or the libido theory in Freud's work makes it possible to isolate the trait that defines the otherness of the sexual to itself. Every stage of Freud's elaboration of the dualism indeed revolves around *a determinate relation of the sexual to death* even before the death drive has appeared as an explicit problem. Each of the couples— sexual drives/ego drives, ego drives/object drives, life drives/death drive— articulate the relation between desire and destruction. Whether it is a matter of the individual's egocentric persistence in the first elaboration of the dualism, the narcissistic enclosure in the second or, of course, the death drive in the third, what is at stake in each case is the relation between life and death. The death drive is present in Freud's theory of the drive from the outset, even if it has yet to be named. For Freud, in other words, the sexual etiology of a psychic disturbance must always begin by taking into account both pleasure and destruction.[11]

As a result, to derive a neurosis from its sexual origin is not to reduce it to the pleasure principle. It is also to confront that which, in the neurosis, proceeds from the tendency toward *pure negation.*

SUBSUMING CEREBRALITY UNDER THE "LIBIDO THEORY," OR THE REVENGE OF PSYCHOANALYSIS

As we have seen, neurologists dispute the existence of a separation between neuronal and psychic energy: neuronal energy is the one and only energy present within all investments, libidinal or otherwise. Wouldn't Freud ask whether the neurologist, unbeknownst to himself, is propounding the thesis of a *new monism,* which, by dismissing the theory of energetic "partition [*Sonderung*]," would deprive itself of the ability to grasp the etiological complexity that actually underlies psychic disturbances?

Because the otherness of the sexual to itself is integral to the etiological significance of sexuality, it implies, for Freud, that every symptom will be evaluated and characterized in terms of the partition of energies—which is to say, *at once in terms of its sexual source and of its tendency to destruction.* In opposition to what one might first suppose, psychoanalysis, far from discounting the causal power of destruction within the genesis of psychic disturbances, very explicitly admits the scope of such a power. Freud simply shows that the operation of destruction never acts alone: within the drive itself, it is always engaged in a struggle, silent or not, with the sexual.

Freud would certainly have insisted on the necessity of *subsuming cerebrality itself under the libido theory*. If the drive is divided from the outset between life and death, the idea of a purely "external," perfectly aleatory, event would be meaningless for the psyche: *everything that happens immediately affects the differential logic of construction and destruction*. If sexuality is always its own other, then, within its very division, it can encompass everything that happens, including the annihilation of the sexual itself. Sexuality is always shadowed by its negative double—which allows it to register events that come to destroy the psyche. Can cerebrality then resist the hermeneutics of the "libido theory"?

Freud would certainly have thought that the contemporary redefinition of cerebrality depends upon uninterrogated epistemological premises. Even if it disregards the dualism of the drive—that is, the originarily cloven structure of sexual causality—the explication of cerebrality would not undermine the etiological privilege of sexuality but, on the contrary, would confirm it.

In fact, within contemporary neurological discourse, it is possible to discern two *confusions*. First, even as it redefines the brain as the site of affects, contemporary neurology would fail to notice that, far from breaking with the psychoanalytic concept of erotogenicity, it actually constitutes the brain as a *privileged erotogenic zone*.

Second, even as it insists upon the fact that cerebral auto-affection can be brutally interrupted by accidental lesions or wounds, contemporary neurology would fail to notice that, far from breaking with the psychoanalytic concept of the death drive, it presents its patients with brain lesions, in their very affective indifference, as examples of a tendency to inertia that returns life to originary organic immobility, a form of retrogression that is characteristic of the death drive.

If Freud were alive today, even if the hybridity of the brain were never discussed, it would appear to him as the *erotic phenomenon* and the *thanatological phenomenon* of our time. The confusion of contemporary neurology would proceed from an oscillation between the "erotic brain" and the "dead brain"—that is, from an inability to decide between the two terms of the dualism of the drive as Freud always conceived them. It is precisely this confusion that would make it necessary to subsume cerebrality under the libido theory.

NEUROLOGICAL CONFUSION BETWEEN THE BRAIN AS EROTOGENIC ZONE
AND THE DEAD BRAIN

In order to dispel this confusion and to subsume cerebrality under the libido theory, it is necessary to formulate two hypotheses that, even as they remain alternatives, are complementary. The first hypothesis consists in supposing that the cool and indifferent attitude of people with brain lesions is *the effect of a narcissistic withdrawal of libido upon the damaged organ.* Indifference can thus be considered—according to the terms of the "second" theory of the dualism of the drive—as a manifestation of the "sexual."

The second hypothesis, which supposes the existence of a death drive at work in the brain, would consider indifference as the expression of a return of the organic to the initial passivity of the inorganic. Emotional indifference could then be considered—according to the terms of the "third" theory of the dualism of the drive—to be a manifestation of the "other of the sexual," that is, of the death drive.

In either case, the emotional indifference or deficit that results from brain damage would accord perfectly with the Freudian conception of psychic eventality as a soldering or a differential structure. No accident whatsoever could fail to be caught within the net of this difference, which would always entwine it within the multiple loops of the conflicts that constitute the drive.

For Freud, the supposed displacement of the psychic causality of sexuality toward cerebrality, in the end, would do nothing more than to rework the effects of the former. The claim that cerebrality constitutes an autonomous cause would only bear witness once again to the etiological power of sexuality.

Brain Damage as Cause of Narcissistic Withdrawal

THE BRAIN AS EROTOGENIC ZONE

What does this mean? The project of constituting of cerebrality as a specific causal character, as well as the subordination of sexual etiology that is its corollary, would hypothetically result from an unconscious operation whose goal is the very opposite of the one that this project claims to pursue: *the constitution of the brain as an erotogenic zone.*

In *Three Essays on the Theory of Sexuality*, Freud defines an "erotogenic zone" as a region of the body liable to become the seat of an excitation of a sexual type. The term can also designate an organ as the source of a "partial sexual drive" (usually translated into English as "component drive").[12] "It is a part of the skin or mucous membrane in which stimuli of a certain sort evoke a feeling of pleasure possessing a particular quality."[13] The erotogenic zones are important because, as an ensemble, they function as a secondary genital apparatus that can even usurp the function of the primary genital apparatus.

Certain regions of the organism, thanks to their peculiar function, are more likely than others to become the seat of excitation. There are "predestined erotogenic zones"—such as the mouth or the anal orifice—that, giving rise to "new sensations," "behave in every respect like a portion of the sexual apparatus."[14] But Freud understands very quickly that any part of the body whatsoever can become an erotogenic zone. He writes: "Any other part of the body can acquire the same susceptibility to stimulation as it possessed by the genitals and can become an erotogenic zone."[15] In fact, "the quality of the stimulus has more to do with producing the pleasurable feeling than the nature of the part of the body concerned."[16]

It is thus possible to conclude, as Freud does in *An Outline of Psychoanalysis*, that "in fact the whole body is an erotogenic zone."[17] In 1914, in "On Narcissism: An Introduction," Freud considerably enlarged the property that he called "erotogenicity [*Erogeneität*]." This property no longer merely pertains to regions of the skin or mucous membrane but also to *internal organs*:

> Let us now, taking any part of the body, describe its activity of sending sexually exciting stimuli to the mind as its "erotogenicity," and let us further reflect that the considerations on which our theory of sexuality was based have long accustomed us to the notion that certain other parts of the body—the "erotogenic zones"—may act as substitutes for the genitals and behave analogously to them. We have then only one more step to take. We can decide to regard erotogenicity as a general characteristic of all organs and may then speak of an increase or decrease of it in a particular part of the body.[18]

Therefore, nothing prohibits us from considering the possibility, even though Freud does not offer this example, that the brain, like any other organ, can become an erotogenic zone. Would it not then be possible to conclude that the contemporary constitution of the brain as the most exposed and most vulnerable sensate zone of all constitutes a *modification of erotogenicity* rather than a subordination of erotogenicity itself to cerebrality? Has not the brain, today, become the very secret of sex—its nongenital intimacy hidden within neuronal economy? "It may well be," Freud writes, "that nothing of considerable importance can occur in the organism without contributing some component to the excitation of the sexual drive."[19] The constitution of cerebrality as an axiological principle regulating a new emotional apparatus, the explication of all the excitable neuronal connections, which can increase or decrease in size and volume, would thus correspond to a redistribution of libidinal economy rather than to its theoretical demotion.

NARCISSISTIC WITHDRAWAL

How could one connect these affirmations with neurological analyses of the psychic impact of brain damage? How might they shed new light upon the problem of emotional indifference? Let us begin with Freud's assertion that: "For every change in the erotogenicity of the organs there might then be a parallel change of libidinal cathexis in the ego."[20] As we can see, every modification of erotogenicity in the body brings about a new configuration of the *narcissism* of the subject.

It is precisely in "On Narcissism: An Introduction" that Freud develops the second version of his theory of the dualism of the drive. From this essay on, the "sexual drives" will no longer be truly opposed to the "ego drives." Freud introduces a "distinction, at the heart of the sexual itself, between the 'ego drives' and the 'object drives': We see also, broadly speaking, an antithesis between ego-libido and object-libido. The more the one is employed, the more the other becomes depleted."[21] Freud is thereby led to conclude that, as the libido is withdrawn from objects, it is not merely depleted but rather displaced on *the ego itself.* The latter becomes at once the object of love and desire: "Libido and ego-interest share the same fate and are . . . indistinguishable from one another."[22]

Accordingly and paradoxically, emotional coolness or indifference toward objects in the external world can very much be considered the outward appearance of a *narcissistic hypercathexis*—that is, the libidinal cathexis of the ego itself. Coolness or disaffection, far from corresponding to a loss of affect, could be nothing other than the mask of a narcissistic passion—that is, desire turned upon the "ego." On Daniel Paul Schreber's narcissistic withdrawal, Freud writes:

> The patient has withdrawn from the people in his environment and from the external world generally the libidinal cathexis which he has hitherto directed on to them. Thus everything has become indifferent and irrelevant to him. . . . The end of the world is the projection of this internal catastrophe; his subjective world has come to an end since this withdrawal of his love from it.[23]

How then is this narcissistic modification, caused by the abandonment of the object and the libidinal cathexis of the ego, linked to "modifications of erotogenicity" in the body? Freud proposes the particularly interesting example of *organic diseases*. Every disease has a tendency *to constitute the sick organ as an erotogenic zone* because of the concentration of the libido upon this organ that isolates it from all other organs or all other objects. This tendency is the narcissistic tendency itself. When there is physical suffering, Freud explains, the ego cathects the sick organ with affect as it withdraws from the external world:

> It is universally known, and we take it as a matter of course, that a person who is tormented by organic pain and discomfort gives up his interest in the things of the external world, in so far as they do not concern his suffering. Closer observation teaches us that he also withdraws his libidinal interest from his love-objects: so long as he suffers, he ceases to love. . . . "Concentrated is his soul," says Wilhelm Busch of the poet suffering from a toothache, "in his molar's narrow hole."[24]

The organ around which the soul is concentrated precisely becomes, in its own way, an erotogenic zone. Libidinal modification constitutes a response to material damage to a specific organic region. What then would prohibit us from considering emotional indifference as the outward manifestation of a *narcissistic scenario*? What would prevent us from concluding that the "soul" of people with brain lesions is concentrated upon "the narrow hole of their brains"?

WAR NEUROSES AND NARCISSISM

The development of Freud's theory of narcissism structures his discourse on traumatic neurosis in general and the war neuroses in particular. The wound has no causal autonomy and it only has the power to engender a neurotic disorder because it is necessarily linked to *erotic modifications*. It is thus impossible to consider any organic damage whatsoever as separate from libidinal redistribution—that is, the narcissistic cathexis of an organ that accompanies its wounding. Accordingly, Freud never fails to link traumatic shock to narcissistic eroticism.

Before developing this point, it is necessary to note that the characteristics of "traumatic neuroses" and "war neuroses" are not quite the same. The former would be classified precisely as "narcissistic neuroses" and the latter as "transference neuroses." In the distinction between these types of neurosis, we discover once again the division between ego drives (since the narcissistic neuroses imply a withdrawal of the ego upon itself) and object drives (since transference constitutes a movement toward the object). At the origin of the transference neuroses, one finds the ego's struggle against attacks from the sexual drive. At the origin of the narcissistic neuroses, which include "pure traumatic neuroses," there is no trace of such a struggle. The violence of the accident and the fright that accompanies it, surging up within the unprepared psyche, cause a turning of the ego within itself, a withdrawal from external objects, and a modification in the erotogenicity of the wounded body.

Even if Freud insists upon distinguishing between the two types of neurosis, he still recognizes that the war neuroses belong to both types. He affirms that "the war neuroses, in so far as they are distinguished from the ordinary neuroses of peacetime by special characteristics, are to be regarded as traumatic neuroses whose occurrence has been made possible by a conflict in the ego."[25] But he immediately specifies that "the war neuroses are only traumatic neuroses, which, as we know, occur in peacetime too after frightening experiences or severe accidents, without any reference to a conflict within the ego."[26]

"Memorandum on the Electrical Treatment of War Neurotics" (1920) attempts to account for the appearance of the enigmatic disorders that are the nervous shocks caused by war. These shocks, Freud asserts, should be placed on the same level as the unexpected accidents or catastrophes that trigger traumatic neuroses:

> There were plenty of patients even in peacetime who, after traumas
> (that is, after frightening and dangerous experiences such as railway accidents,
> etc.) exhibited severe disturbances in their mental life and their nervous

activity, without physicians having reached an agreed judgment on these states. . . . The war that has recently ended produced and brought under observation an immense number of these traumatic cases.[27]

The war neurotic is thus sick in two ways. He suffers both from a transference neurosis and from an accidental trauma. The first disturbance, as we have shown, stems from a conflict in the ego.[28] In the majority of war neuroses, the conflict "is between the soldier's old peaceful ego and his new warlike one."[29] It is this conflict that justifies fear—the urge to flee or desert. The new warlike ego is akin to an enemy and thus appears in the form of a threat that must be expunged at all costs. The danger that is "embodied" in the soldier's ego is thus analogous to the threat that the libido represents for the ego in peacetime.[30] The split between the peaceful ego and the warlike ego that occurs in real military conflict reproduces the conflict between the ego drives and the sexual drives.

The second type of disturbance, which results from a traumatic element, owes nothing to this conflictual structure: "Apart from this, the war neuroses are only traumatic neuroses, which, as we know, occur in peacetime too after frightening experiences or severe accidents, without any reference to a conflict in the ego."[31] Nonetheless, what's at stake here is not the pure intervention of an *Ereignis* beyond any possible preparation or staging. Freud's response is without appeal: even if the traumatic element does not stem from a conflict in the ego, it remains linked to this conflict, and it must refer to sexuality as its principle of explanation.

Freud immediately dismisses any suggestion that an organic cause could have etiological autonomy: "The terrible war which has just ended gave rise to a great number of illnesses of this kind, but it at least put an end to attribute the cause of the disorder to organic lesions of the nervous system brought about by mechanical force."[32] And in "Memorandum on the Electrical Treatment of War Neurotics," he deplores the fact that doctors could not discern the true causes of war neurosis:

> Some supposed that with such patients it was a question of severe injuries to the nervous system, similar to the haemorrhages and inflammations occurring in non-traumatic illnesses. And when anatomical examination failed to establish such processes, they nevertheless maintained their belief that finer changes in the tissues were the cause of the symptoms observed. They therefore classed these traumatic cases among the organic diseases.[33]

Fortunately, these same doctors have gradually begun to adopt psychoanalytic insight:

> The great majority of physicians no longer believe that the so-called "war-neurotics" are ill as a result of tangible organic injuries to the nervous system, and the more clear-sighted among them have already decided, instead of using the indefinite description of a "functional change," to introduce the unambiguous term "mental change."[34]

Of course, "mental," here, is opposed to "cerebral." Brain damage or head wounds cannot give rise to a neurosis.

One last passage confirms this point:

> Although the war neuroses manifested themselves for the most part as motor disturbances—tremors and paralyses—and although it was plausible to suppose that such a gross impact as that produced by the concussion due to the explosion of a shell near by or to being buried by a fall of earth would lead to gross mechanical effects, observations were nevertheless made which left no doubt as to the psychical nature of the causation of the so-called war neuroses. How could this be disputed when the same symptoms appeared behind the Front as well, far from the horrors of war, or immediately after a return from leave? The physicians [i.e. psychoanalysts] were therefore led to regard war neurotics in a similar light to the nervous subjects of peacetime.[35]

How should we understand this refusal to accord "patent" wounds the status of *determining causes*? And what happens to the "accidental" element in the constitution of neurotic disturbances? There is a ready answer to these questions: when trauma occurs, *it serves as an escape from an internal conflict.* The patient takes refuge in neurosis not in order to avoid *the war being waged at the front* but rather *the war raging within himself.* Freud could not be clearer: "It would be equally true to say that the old ego is protecting itself from a mortal danger by taking flight into a traumatic neurosis or to say that it is defending itself against the new ego which it sees as threatening its life."[36] This point is confirmed in *Beyond the Pleasure Principle*: "I have argued elsewhere that 'war neuroses' (in so far as that term implies something more than a reference to the circumstances of the illness's onset) may well be traumatic neuroses which have been facilitated by a conflict in the ego."[37]

Traumatic neurosis, unlike transference neurosis, does not stem from an internal conflict, but it does provide a *means of escape* from such a conflict.

In this sense, it remains utterly dependent upon internal conflict—which explains why traumatic neurosis, exactly like transference neurosis, has a *sexual* etiology. In "Introduction to *Psychoanalysis and the War Neuroses*," Freud admits that it is difficult to explain the traumatic element in terms of the "sexual etiology of the neuroses" and of the "libido theory." He asserts:

> The theory of the sexual etiology of the neuroses, or, as we prefer to say, the libido theory of the neuroses, was originally put forward only in relation to the transference neuroses of peacetime and is easy to demonstrate in their case by the use of the technique of analysis. . . . But the traumatic neuroses of peacetime have always been regarded as the most refractory material of all in this respect; so that the emergence of the war neuroses could not introduce any new factor into the situation that already existed.[38]

However, Freud's new elaboration of the dualism of the drive, in "On Narcissism: An Introduction," made it possible to confirm the sexual etiology of the neuroses. Traumatic neuroses, much like "ordinary dementia praecox, a paranoia, or a melancholia" emerge as illnesses that result from the narcissistic withdrawal of libido. The sexual etiology of the war neuroses thus remains untouched. In the narcissistic neuroses, as we have seen, the "sick man withdraws his libidinal cathexes back upon his own ego."[39] An organic illness resulting from the impact of a wound or a shock upon the distribution of libido would thus lead the ego to withdraw its interest from objects in order to concentrate upon itself. Accordingly, illness implies a "narcissistic withdrawal of the positions of the libido on to the subject's own self."[40] People who have been traumatized in the experience of war thus behave exactly like narcissistic neurotics. On a psychic level, the war wound does not introduce anything more or, above all, anything other than organic illness or shocks that occur in "peacetime."

WOUNDS AGAINST NEUROSIS

It is now possible to explain an enigmatic assertion from *Beyond the Pleasure Principle*:

> In the case of ordinary traumatic neuroses two characteristics emerge prominently: first, that the chief weight in their causation seems to rest upon the factor of surprise, of fright; and secondly, that a wound or injury inflicted simultaneously works as a rule *against* the development of a neurosis.[41]

Let us consider the first. For Freud, as we know, *every shock produces sexual excitement*. In the case of catastrophic accidents, sexual excitement occurs without warning, since the psychic apparatus could not anticipate it nor produce the anxiety that would have been necessary for such preparation. Fright (*Schreck*) is the consequence of this absence of anxiety: "The mechanical violence of the trauma would liberate a quantity of sexual excitation which, owing to the lack of preparation for anxiety, would have a traumatic effect."[42] The chain of events unfolds in the following order: (1) the accident occurs, (2) sexual excitation occurs with it, and then (3), with this unforeseen excitation, there is fright. The "shield against stimuli" is "pierced" (in the proper sense of *titrosko*). The result is "ordinary traumatic neurosis."

Let us now consider the second. Were a neurosis to be accompanied by a physical lesion, *it would disappear as such*. Indeed, Freud even affirms that an organic wound replaces neurosis and thereby, in a sense, *cures* it:

> It is also well known, though the libido theory has not yet made sufficient use of the fact, that such severe disorders in the distribution of libido as melancholia are temporarily brought to an end by intercurrent organic illness, and indeed that even a fully developed condition of dementia praecox is capable of temporary remission in these same circumstances.[43]

Because bodily wounds provoke a narcissistic hypercathexis, they make it possible to neutralize energetic panic, to discipline the excess caused by trauma, and to gather it within a unity, focusing the scattered anxiogenic libido liberated by the accident upon the wounded organ. The wound is thus an agent of psychic reparation and recovery in the wake of a traumatic effraction.

In this sense, *the wound is the means whereby trauma heals itself*. The wound, paradoxically, is already a scar. For Freud, physical damage—in particular, lesions of the nervous system—functions to repair the psyche.[44]

Accordingly, in cases of war neuroses no less than in cases of traumatic neuroses, the "libido theory" retains its etiological primacy either by deriving neurosis from a conflict of the ego with itself or by concluding that the concentration of the libido upon the wound conjures the threat of neurosis. The auto-eroticized wound fends off the return of an ego-conflict. Sexuality thus intervenes itself in every case in order to activate or to efface the psychic effects of organic damage, especially brain damage.

Brain Damage and the Death Drive

I now turn to an examination of the second hypothesis that concerns the work of the death drive in the brain. According to this hypothesis—which, as always, adheres to the tenets of the libido theory—the indifference or disaffection of people with brain lesions would only be manifestations of the death drive rather than reactions of a new type to destructive events coming from outside.

To envisage the emotional deficit of people with brain damage as a narcissistic phenomenon is to envisage it as a manifestation of the sexual drive. The time has come to attempt an interpretation of this same deficit in terms of the death drive.

The distinction between these two drives is not a simple alternative. On one hand, as Freud often emphasizes, the two types of drive—the life drives (which include the sexual drive) and the death drive—are always fused within one another. If "defusions" remain possible, they have to begin from the original "mixture."[45] On the other hand, in *Beyond the Pleasure Principle*, before Freud groups all types of sexual drive together with the life drives, he considers the possibility that the ("narcissistic") ego drives are also expressions of the death drive, in the sense that their aim is to preserve the isolated individual rather than to perpetuate the species.[46] These two hypotheses therefore do not exclude one another.

To interpret the behavior of people with brain lesions as the expression of the death drive is to interpret their indifference as a manner of *letting oneself die*. Brain damage, once again, would not be the *determining* cause of disaffection but merely the cause of a defusion of the drives that delivers the organism unto the internal imperative to destroy itself. It would thus not be possible to think emotional disaffection outside a process of self-destruction that can only be understood in terms of the "libido theory" rather than as a purely neurological phenomenon.

"THE ORGANISM WISHES TO DIE ONLY IN ITS OWN FASHION"

In *Beyond the Pleasure Principle*, Freud establishes that all drives are conservative in that they seek to restore an earlier state of things. When he undertakes the analysis of the posttraumatic behavior that leads patients to reproduce, in their dreams, the scene of their accidents, Freud encounters the existence of a "compulsion to repeat" (*Wiederholungszwang*). The latter highlights the regressive character of the drive:

> But how is the predicate of being "driven" related to the compulsion to repeat? At this point we cannot escape a suspicion that we may have come upon the track of a universal attribute of the drives and perhaps of organic life in general which has not hitherto been clearly recognized or at least not explicitly stressed. It seems, then, that a drive is an urge inherent in organic life to restore an earlier state of things which the living entity has been obliged to abandon under the pressure of external disturbing forces.[47]

Contrary to what one might initially believe, the drive does not push toward change but rather toward the past. This is how Freud comes to posit the strange law according to which *the aim of all life is death*. Not simply the cessation of life, but rather its end (*Ende*), in the sense of finality, the terminal point that life pursues. Indeed, "the elementary living entity would from its very beginning have had no wish to change; if conditions remained the same, it would do no more than constantly repeat the same course of life."[48] Repetition thus opens toward "an *old* state of things, an initial state from which the living entity has at one time or other departed and to which it is striving to return by the circuitous paths along which its development leads."[49] This anterior state is death or "organic passivity." Accordingly, because "inanimate things existed before living ones," it is quite legitimate to uphold that "the aim of all life is death."[50]

It is thus an *external* force that comes to disorient, for a time, the immanent movement toward death and thus to preserve life. But this force is only a "detour" on the path toward death. Indeed, this path itself is merely a further detour![51] The longer life becomes in superior living entities, the more the death drive must *impose a detour upon the detour* of this lengthening in order to achieve its own aim and to lead life *backwards*. The specificity of the conservative character of the death drive with respect to all the other drives consists in this aim to "return to the inanimate state."[52]

Freud thus distinguishes the life drives and the death drive in the following terms.

> [The life drives] are conservative in the same sense as other drives in that they bring back earlier states of living substance. . . . [But] they operate against the purpose of the other drives, which lead, by reason of their function, to death; and this fact indicates that there is an opposition between them and the

other drives, an opposition whose importance was long ago recognized by the theory of the neuroses. It is as though the life of the organism moved with a vacillating rhythm. One group of drives rushes forward so as to reach the final aim of life as swiftly as possible; but when a particular stage in the advance has been reached, the other group jerks back to a certain point to make a fresh start and to prolong the journey.[53]

Life thus appears as the specific path that the individual follows toward his own death. "The organism," Freud writes, "only wishes to die in its own fashion."[54] The life drives, in reality, constitute the opening of one's *own* death. By fashioning a singular path, they adhere to the immanent character of the end. Even the "drives of self-preservation," grouped with the life drives, and thus with the sexual drives, "are component drives whose function is to assure that the organism shall follow its own path to death, and to ward off any possible ways of returning to inorganic existence other than those which are immanent to the organism itself."[55]

The event of death or the accident of death is only possible on the basis of the instinctual immanence of death.[56] In *Difference and Repetition*, Deleuze underscores this *double status of death*:

> Every death is double. . . . Freud suggested the following hypothesis: the organism wants to die, but to die in its own way, so that real death always presents itself as a foreshortening, as possessing an accidental, violent and external character which is anathema to the internal will-to-die. There is a necessary non-correspondence between death as an empirical event and death as a "drive" or transcendental instance. . . . Desired from within, death always comes from without in a passive and accidental form.
> Suicide is an attempt to make the two incommensurable faces coincide or correspond.[57]

Would it not be possible to conclude that psychic destruction is itself also necessarily double, at once immanent and accidental, not merely one or the other? If this were the case, it would become impossible to think the catastrophic event without convoking a "being-toward-death."

The "imperishable" character of psychic life thus has a signification entirely other than Freud supposes. Far from being a manifestation of indestructibility, it becomes one of the expressions of the death drive, of its own compulsion to repeat, or of the eternal return of the identical. *The imperishable is death itself.*

SELF-DESTRUCTION

In this sense, a lesion, a shock, a trauma, or a catastrophic event would at once trigger, confirm, and accelerate *the drive of self-destruction* present in every living individual:

> The one set of drives, which work essentially in silence, would be those which follow the aim of leading the living creature to death and therefore deserve to be called the "death drives"; these would be directed outwards as the result of the drives themselves as destructive or aggressive impulses.[58]

The death drive manifests itself as a propensity to destruction and aggression (the desire "to destroy and kill"[59]) when it is directed outwards. As Freud explains in *The Ego and the Id*,

> It appears that, as a result of the combination of unicellular organisms into multicellular forms of life, the death drive of the single cell can successfully be neutralized and the destructive impulses be diverted on to the external world through the instrumentality of a special organ. This special organ would seem to be the muscular apparatus; and the death drive would thus seem to express itself—though probably only in part—as a drive of destruction directed against the external world and other organisms.[60]

Accordingly, nothing prohibits us from concluding that the drive of destruction and the drive of aggression are in fact "diverted" expressions of indifference or organic inertia. A few years later, Freud will declare, in *Why War*, that "the death drive turns into the destructive drive when, with the help of special organs, it is directed outwards, onto objects."[61]

Once again, is not the indifference or disaffection of people with brain lesions simply the effect of multiple detours within this same tendency, within the same conflict between life and death—a tendency and a conflict that exceed the narrow framework of neuronal functioning to the extent that they *govern* it?

A passage from *The Ego and the Id* is eloquent in this respect:

> We suspect that the epileptic fit is a product and indication of an instinctual defusion; and we come to understand that instinctual defusion and the marked emergence of the death drive call for particular consideration among the effects of some severe neuroses.[62]

If an epileptic attack is a phenomenon of the death drive, why not enlarge the list of "severe neuroses" to include the entire set of "neuropathological disorders" and conclude that they are, too, consequences of instinctual defusion?

QUESTIONS

Objections continue to pile up against the initial affirmation that there are two separate evental regimes: cerebrality and sexuality. How is it possible, after all of these developments, that trauma, the accident, or the unforeseen catastrophe plays a *determining* role within the psyche? How is it still possible to affirm that a creative destruction is at work within the formation of barren, unprecedented, or unrecognizable identities? How is it possible—because death itself is a "soldering" that reveals the conjunction between an unforeseeable event and the metabolism of drive?

SEVEN

Separation, Death, the Thing, Freud, Lacan, and the Missed Encounter

> The frightening unknown on the other side of the line is that which in man we call the unconscious, that is to say the memory of those things he forgets . . . those things in connection with which everything is arranged so that he doesn't think about them, i.e. stench and corruption that always yawn like an abyss.
>
> — JACQUES LACAN, *Seminar VII: The Ethics of Psychoanalysis*

Methodological Clarification

If it still remains possible to constitute cerebrality as a specific regime of events, if it remains possible to show that trauma—as unexpected accident or unforeseen catastrophe—possesses a *determining* and not merely a *triggering* power within the psyche, this possibility could only be elaborated at the very heart of Freud's thinking of danger, destruction, and the annihilation of the psyche. It would thus be necessary to enter as deeply as possible into this thinking and to show the precise way in which the place of cerebrality opens up within the powerful economy of the death drive.

Once again, the confrontation that I am staging here is not simple. As my argument advances, it becomes even clearer that it is impossible to naïvely oppose the neurological discourse on the psychic impact of brain lesions to the theory of neurotic predisposition. Freud, we have shown, accords a fundamental role to the factor of surprise, fright, and the psyche's lack of preparation for the external accident or danger.

The cloven structure of sexuality, the "screen" of the "libido theory" (according to which sexuality is also—and perhaps before all—encompasses the other of the sexual or the otherness of sexuality to itself) enlarges the signification of etiology. Indeed, by virtue of this structure, sexual etiology accords a fundamental place to *mortiferous* eventality that cannot be separated from sexuality as such even as it is foreign to sexual *life*.

The meaning of Freud's refusal of any instinctual monism ultimately lies in the recognition that there are *two sources of events at the very heart of the same causal character*. Freud readily admits that the psyche is exposed to unforeseeable catastrophes of destructive events. The dual organization of the concept of sexuality makes it possible to account for the *constitution* as well as the *destruction of the psyche*.

The confrontation between sexuality and cerebrality, therefore, should not become an exposition that contrasts between two types of discourse: a neurological discourse that would insist on the psyche's constitutive vulnerability and a psychoanalytic discourse that would insist on its imperishable character. Contrary to what one might be led to suppose, such a confrontation distinguishes *two apparently similar but, in fact, radically opposed concepts of destruction*.

I will now examine *the status of the psychoanalytic hypothesis of an annihilation of the psyche*—or rather, the hypothesis of *absolute danger*. It is necessary to begin from this hypothesis in order to show what remains unquestioned within it. Only such a critical opening can give cerebrality its *chance*. Where and how are these two visions of destruction distinguished from one another? This is the question that the present chapter seeks to address.

The Freudian Idea of Death as Separation from Self

THE UNCONSCIOUS AS THE ORIGINARY RELATION BETWEEN THE PSYCHE AND ITS OWN DESTRUCTION

Freud affirms that the unconscious knows no negation, time, or death.[1] He insists on the perennial longevity of unconscious structures, on the force of the phylogenetic inheritance of pasts not lived. Nonetheless, for Freud, this very perenniality bears witness to the fact that *the unconscious is nothing other than the form of the originary relation between the psyche and its own destruction.* As Deleuze quite rightly underscores, "Freud supposes the unconscious to be ignorant of three important things: Death, Time, and No. Yet it is a question only of death, time, and no in the unconscious."[2] Everything that making and letting die might mean for the psyche, in fact, is elaborated at great length in a work that never stops referring to annihilation, to an Ego capable of letting itself perish as it gives into the anxiety about death that occurs both "as a reaction to an external danger (*aüssere Gefahr*) and as an internal process (*innerer Vorgang*), as for instance in melancholia."[3] To do justice to Freud, one must show that, for him, danger comes as much from *outside* as *inside* the psyche. He asserts:

> The fear of death in melancholia only admits of one explanation: that the ego gives itself up because it feels itself hated and persecuted by the super-ego, instead of loved. . . . But, when the ego finds itself in an excessive real danger [*realen Gefahr*] which it believes itself unable to overcome by its own strength, it is bound to draw the same conclusion. It sees itself deserted by all protecting forces and lets itself die [*lässt sich sterben*].[4]

Freud in no way minimizes the importance of external threats or perils. It is thus not a matter of contesting the existence, in Freud, of an idea of psychic destruction but rather of examining and discussing *the unconscious status of this idea and the role that it plays within the psyche itself.* How does the psyche apprehend its own end?

"REAL" DANGER AND THE CUT: CASTRATION, PUNISHMENT, AND BIRTH

For Freud, the relation of the psyche to its own disappearance is envisaged as a separation of the psyche from itself. Separation from self is the psychic phenomenon of mortality. For Freud, the anticipation of death is lived as the ego's farewell to itself. Dying signifies taking leave of oneself. Anxiety, as a reaction to danger, is then always fundamentally *separation anxiety, the affect of cutting as such.* These precise definitions will become the focus of the controversy with neurology.

But let us begin with the Freudian conception of anxiety: Whether it is the affective expression of internal instinctual danger or whether, on the contrary, it is a reaction to an external threat, anxiety remains the affect of distancing or tearing apart. From his *Papers on Metapsychology*, where Freud contends that anxiety is *produced by* repression and is a reaction to internal instinctual peril, to *Inhibitions, Symptoms, and Anxiety*, where anxiety becomes the direct reaction to a "real" external danger and the *cause* of repression, the basis of the Freudian idea of the destruction of the psyche remains the phenomenon of separation from self.

Certainly, the theoretical trajectory from one work (*Papers on Metapsychology*) to the other (*Inhibitions, Symptoms, and Anxiety*) is remarkably complex and the conclusions in each work even appear to contradict one another. In *Papers on Metapsychology*, in fact, anxiety appears as a consequence of repression while, later, repression appears to be a creation of anxiety. In *New Introductory Lectures on Psychoanalysis*, Freud addresses this "inversion" of the relation between anxiety and repression: "It was not repression that created the anxiety . . . it was anxiety that made the repression."[5] Importantly, this inversion results from a new attempt to take *danger* into account. At this point in Freud's thinking, it is no longer internal instinctual danger that appears as the prime form of all danger; for, he shows that this instinctual danger itself is only the internalization of a real material danger: "But what kind of anxiety can it have been? Only anxiety in the face of a threatening external danger—that is to say, a real anxiety (*Realangst*)."[6] Indeed: "It must be confessed that we were not prepared to find that internal instinctual danger would turn out to be a determinant and preparation for an external, real, situation of danger."[7]

Freud then affirms that every "instinctual situation which is feared goes back ultimately to an external situation of danger."[8] In fact, he pursues, "neurotic anxiety has changed in our hands into real anxiety, into fear of particular external situations of danger."[9] Freud will no longer be concerned, as he was in *Papers on Metapsychology*, with the psychic importance of "how a neurotic anxiety is changed into an apparently real one,"[10] but, on the contrary, with the priority of real anxiety itself as a response to an actual external danger.

What, then, is this danger that is more originary than internal danger? It is precisely the danger of separation, which, at this stage of Freud's analysis, appears in the three principal forms of *birth*, *punishment*, and *castration*. Beginning from the last in the series, Freud explains the order in which these forms are derived. The "punishment of castration," the "loss," for the little boy, "of his member," appears as a "real danger that the child fears as a result of his love for his mother." Freud clarifies:

You will of course object that after all that is not a real danger. Our boys are not castrated because they are in love with their mothers during the phase of the Oedipus complex. But the matter cannot be dismissed so simply. Above all, it is not a question of whether castration is really carried out; what is decisive is that the danger is one that threatens from outside and that the child believes in it.[11]

Castration anxiety (the third form of separation) is itself a substitute for the fear of punishment—punishment by the mother who threatens to withdraw her love for the child (the second form of separation); and this punishment anxiety, in turn, is the expression of an even older anxiety linked to the trauma of birth (the first form of separation):

> If a mother is absent or has withdrawn her love from her child, it is no longer sure of the satisfaction of its needs and is perhaps exposed to the most distressing feelings of tension. Do not reject the idea that these determinants of anxiety may at bottom repeat the situation of the original anxiety at birth, which, to be sure, also represented a separation from the mother. Indeed, if you follow a train of thought suggested by Ferenczi, you may add the fear of castration to this series, for a loss of the male organ results in an inability to unite once more with the mother (or a substitute for her) in the sexual act. I may mention to you incidentally that the very frequent fantasy of returning to the mother's womb is a substitute for this wish to copulate.[12]

Finally, Freud turns to the work of Otto Rank, who "has the merit of having expressly emphasized the significance of the act of birth and of separation from the mother."[13] And Rank leads him to take into account the fact of "helplessness [*Hilflosigkeit*]," the distress of the newborn engendered by its state of immaturity and absolute dependence. As the child grows up, this distress continues to develop and his separation anxiety takes a different form at every age:

> The danger of psychical helplessness fits the stage of the ego's early immaturity; the danger of loss of an object (or loss of love) fits the lack of self-sufficiency in the first years of childhood; the danger of being castrated fits the phallic phase; and finally the fear of the super-ego, which assumes a special position, fits the period of latency.[14]

These three "dangers" of separation—birth, punishment, and castration— are analyzed in the same fashion in *Inhibitions, Symptoms, and Anxiety*. In this work, Freud more clearly distances himself from Rank, showing that his hypothesis of "birth trauma" is "unfounded and extremely improbable."[15] Nonetheless, separation always appears as the predominant factor when the psyche is exposed to danger and birth remains one of the most striking phenomena of this separation. Separation is inflected into what Lacan calls the "five forms of loss, *Verlust*, that Freud designates in *Inhibitions, Symptoms, and Anxiety*"[16]—that is, the gradation birth-castration-loss of love-punishment-exclusion:

> We have already traced the change of that context from the loss of the mother as an object to castration. The next change is caused by the power of the superego. With the depersonalization of the parental agency from which castration was feared, the danger becomes less defined. Castration anxiety develops into moral anxiety—social anxiety . . . "separation and expulsion from the horde" . . . The final transformation which the fear of the superego undergoes it, it seems to me, the fear of death (or fear for life) which is a fear of the superego projected onto the powers of destiny.[17]

It emerges clearly from these analyses that the situation of real danger is finally much older than instinctual danger, which then appears as its trace or memory. Phobia, for example, is always derived from this primary danger, of which it is only a substitute. Returning to the case of "Little Hans," Freud modifies his original conclusions and writes:

> As soon as the ego recognizes the danger of castration it gives the signal of anxiety and inhibits through the pleasure-unpleasure agency . . . the impending cathectic process in the id. At the same time, the phobia is formed. And now the castration anxiety is directed to a different object and expressed in a distorted form, so that the patient is afraid, not of being castrated by his father, but of being bitten by a horse or devoured by a wolf. . . . There is no need to be afraid of being castrated by a father who is not there. On the other hand one cannot get rid of a father; he can appear wherever he chooses. But if he is replaced by an animal, all one has to do is avoid the sight of it—that is, its presence—in order to be free from danger and anxiety. "Little Hans," there-fore, imposed a restriction upon his ego. He produced the inhibition of not leaving the house, so as not to come across any horses.[18]

Repression becomes actual suppression rather than production of anxiety:

> The ego notices that the satisfaction of an emerging instinctual demand would conjure up one of the well-remembered situations of danger. This instinctual cathexis must therefore be somehow suppressed, stopped, made powerless. . . . With this the automatism of the pleasure-unpleasure principle is brought into

operation and now carries out the repression of the dangerous instinctual impulse.[19]

As his theory of anxiety evolves, Freud seems to elevate the process of individual separation—produced within the drive itself by a repression that divides representation and affect—to the status of a transcendental experience: originary separation.

FEAR OF DEATH

All of the previously evoked dangers—birth, loss of love, and castration— manifest the fact that the psyche can only represent its own annihilation in the mode of a *cut* or a *dissociation from self*. Indeed, in the unconscious *there is no meaningful content associated with the concept of one's own death*. "My" death remains unrepresentable: "for death is an abstract concept with a negative content for which no unconscious correlative can be found."[20] Or: "But the unconscious seems to contain nothing that could give any content to our concept of the annihilation of life. . . . Nothing resembling death can ever have been experienced; or, if it has, as in fainting, it has left no observable traces behind."[21] Separation alone can provide an image of what would otherwise remain pure abstraction.

Just as "castration can be pictured on the basis of the daily experience of the faeces being separated from the body or on the basis of losing the mother's breast at weaning,"[22] death, in turn, can only be represented through the (not directly representable) experience of castration. The only sensible given that allows for the phenomenal translation of destruction is loss or the cut:

> I am therefore inclined to adhere to the view that the fear of death should be regarded as analogous to the fear of castration and that the situation to which the ego is reacting is one of being abandoned by the protecting superego—the powers of destiny—so that it no longer has any safeguard against all the dangers that surround it.[23]

In the same way that everyday physiological separations "represent" castration, castration itself "represents" the ego's farewell to itself. Through such representation, death only becomes one's own, "my death," to the extent that it appears as the process whereby psychic instances dissociate from one another—as if the ego gave itself the slip, spent its time preparing for and negotiating its own departure, a departure that will have been intimated, throughout its life, by "constantly repeated object-losses."[24]

Freud's analysis of the motif of the double in *The Uncanny* confirms this point. The uncanny, the strange, or *unheimlich*, is "everything that ought to have remained secret and hidden but has come to light."[25] That which suddenly emerges from its hiding place is also "dangerous." The remainder of the analysis, with its very beautiful reading of *The Sandman*, will increasingly identify the *unheimlich* with the motif of the double. The instance that emerges from hiding is the very intimacy (*Heimlichkeit*) of the ego, its inside, which, by taking leave of its "box," prefigures the moment of fatal separation. The ego prefigures death as the avowal that a part of itself—the double of the other—has relinquished life and departed. The double is the ego that is ready to leave the other, to go ahead into nothingness. In *The Ego and the Id*, Freud writes: "It would seem that the mechanism of the fear of death can only be that the ego relinquishes its narcissistic libidinal cathexis in a very large measure—that is, that it gives itself up, just as it gives up some *external* object in other cases in which it feels anxiety."[26] Culture has transformed the double—long considered the immortal part of the self, as its incorruptible "soul"—into the messenger of death:

> For the "double" was originally an insurance against the destruction of the ego, an "energetic denial of the power of death," as Rank says; and probably the "immortal" soul was the first "double" of the body. . . . Such ideas, however, have sprung from the soil of unbounded self-love, from the primary narcissism which dominates the mind of the child and of primitive man. But when this stage has been surmounted, the "double" reverses its aspect. From having been an assurance of immortality, it becomes the uncanny harbinger of death.[27]

The ghostly double, the dead man "becomes the enemy of his survivor and seeks to carry him off to share his new life with him."[28] The ego doubles itself, and this scission opens the psyche to the horizon of its own disappearance.

CLARIFICATION OF THE RELATION BETWEEN SEXUAL ETIOLOGY AND THE FEAR OF DEATH

This fundamental point makes it possible to return to the complex—that is, cloven—structure of sexual etiology. I asserted that the otherness of sexuality to itself solidifies rather than threatening the unity of its causal power. However, we can now see that there is an indissoluble relation between sexuality and death to the extent that death is prefigured in castration. Haunted by the fear of loss—loss of a member, a part of the body, or loss of self—sexuality is itself structured by the anticipation of separation and death. The fear of death is thus not an affect that draws the psyche beyond or away from sexual causality, but rather is the immanent manifestation of the *distinctive feature of both sexuality and death: separation.*

In *Inhibitions, Symptoms, and Anxiety*, Freud returns to the question of traumatic neurosis precisely in order to highlight the role of sexual etiology within it. In the eyes of the detractors of psychoanalysis, traumatic neurosis seems utterly foreign to such etiology. How could trauma—in all its suddenness, with its unforeseen character—have anything to do with sexuality? How could it awaken castration anxiety? Reviewing the arguments of his adversaries, Freud writes:

> If anxiety is the reaction of the ego to danger, we shall be tempted to regard the traumatic neuroses, which so often follow upon a narrow escape from death, as a direct result of a fear of death (or a fear *for* life) and to dismiss from our minds the question of castration and the dependent relations of the ego. Most of those who observed the traumatic neuroses that occurred during the last war took this line, and triumphantly announced that proof was now forthcoming that a threat to the drive of self-preservation could by itself produce a neurosis without any admixture of sexual factors and without requiring any of the complicated hypotheses of psychoanalysis.[29]

Nonetheless, responds Freud, such arguments represent a misunderstanding of castration anxiety! In fact, this anxiety does not primarily represent the loss of a specific object but rather the *indeterminate threat of a cut.* Accordingly, everything that places life in danger necessarily occurs as a blade that threatens to separate the ego from itself, to cut it to the quick. This is the sense in which the danger of death corresponds to the danger of castration. Every self-preservative drive is an originary response to the possibility of separation. Narcissism itself is only thinkable as a reaction to the possibility of cutting. It thus becomes obvious that the only satisfying manner to explain traumatic neuroses is in terms of sexual etiology. Freud pursues: traumatic neurosis does not "contradict the etiological importance of sexuality" to the extent that "any such contradiction has long since been disposed of by the introduction of the concept of narcissism, which brings the libidinal cathexis of the ego in line with the cathexes of objects and emphasizes the libidinal character of the drive of self-preservation."[30]

For Freud, sexual etiology is thus perfectly suited to account for the fear of death. Fundamentally, it is the cut—the fantasy or anticipation thereof—that opens the psyche to the horizon of its own relation to itself, to the way in which it can *see itself die by doubling itself*. Mortiferous separation from self is the very origin of the speculation or reflection whereby the ego takes itself as an object. The birth of the superego is thus inseparable from this process of self-observation:

> A special agency is slowly formed [in the ego], which is able to stand
> over against the rest of the ego, which has the function of observing and
> criticizing. . . . The fact that an agency of this kind exists, which is able to
> treat the rest of the ego like an object—the fact, that is, that man is capable
> of self-observation—renders it possible to invest the old idea of a "double"
> with a new meaning.[31]

This horizon of observation, which opens up thanks to the way in which process of mortal doubling distances the ego, pertains to the structure of anticipation that every form of anxiety—internal or external—has in common. By the same token, it is the apparatus of psychic openness to all types of events and accidents. Whether it is "materially real" or "psychically real," the event comes to affect a structure of anticipation elaborated on the basis of the originary possibility of leaving oneself behind.

The Indestructible Character of Annihilation

The distinction between *Ereignisse* (external events) and *Erlebnisse* (lived events) breaks down a bit further. Every accident, whether "endogenous" or "exogenous," affects the structure of anticipation opened by the separation of the ego from itself. This *first separation* is, according to Freud, the *first trauma, a "real" event without being*. Originary separation takes place without actually happening. The first trauma is *at once Ereignis and Erlebnis without being more one than the other*.

Freud goes so far as to risk the hypothesis of a primitive trauma that occurs to the species as a whole—to everyone and no one. This would be a "particular important event [*ein gewissen bedeutungsvoll Ereignis*], incorporated by inheritance—something that may thus be likened to an individually acquired hysterical attack."³² However, it is obviously impossible to know whether or not such an event has really occurred, whether it ever happened *to* the psyche or whether it is a creation *of* the psyche.

In *Overview of the Transference Neuroses*, Freud, developing Ferenczi's idea, advances the hypothesis that this trauma dates from the Ice Age:

> Our first hypothesis would thus maintain that mankind, under the influence
> of the privations that the encroaching Ice Age imposed upon it, has become
> generally anxious [*ängstlich*]. The hitherto predominantly friendly outside
> world, which bestowed every satisfaction, transformed itself into a mass
> of threatening perils. There had been good reason for real anxiety about
> everything new.³³

Whether this experience of anxiety is the phylogenetic transmission of a distressing memory or whether it is, on the contrary, the unexpected upsurge of sudden anxiety, and regardless of the "long dispute over whether real anxiety . . . is more originary,"³⁴ what matters in the end is the fact that anxiety takes place within a structure of anticipation that exists before it and that is the very form of the unconscious.

In one of the addenda to *Inhibitions, Symptoms, and Anxiety*, entitled "Supplementary Remarks on Anxiety," Freud insists on the proximity of anxiety and anticipation. What is danger? Is it a matter of the surprise that it provokes the "first time" that it occurs or rather of the dread of its repetition? Is it a matter of newness or of recurrence? It is not possible to separate the one from the other since anticipation is always constituted of both instances at once:

> The individual will have made an important advance in his capacity for
> self-preservation if he can foresee and expect [*erwartet*] a trauma situation of
> this kind which entails helplessness, instead of simply waiting for it to happen
> [*abgewartet*]. Let us call a situation of this kind which contains the determinant
> for such an expectation a danger-situation. It is in this situation that the signal
> of anxiety is given. The signal announces: "The present situation reminds me
> of one of the traumatic experiences I have had before. Therefore I will
> anticipate the trauma and behave as though it had already come, which there is
> yet time to turn it aside." Anxiety is therefore on the one hand an expectation
> of a trauma, and on the other a repetition of it in a mitigated form.³⁵

In the beginning, it might be that trauma was "passively experienced," that it was purely and simply undergone, and that the psyche bears the trace of this experience, a trace that is always liable to be reactivated upon the return of any threatening situation. At the same time, it is now quite clear that the distinction between the experienced and the not experienced, the first time and the repetition, anticipation and recollection, breaks down to such an extent that these oppositions necessarily belong to one and the same structure.

This breaking down of distinctions also affects the distinction between external and internal danger:

> In relation to the traumatic situation, in which the subject is helpless, external and internal dangers, real dangers and instinctual demands converge. Whether the ego is suffering from a pain which will not stop or experienced an accumulation of instinctual needs which cannot obtain satisfaction, the economic situation is the same, and the motor helplessness of the ego finds expression in psychical helplessness.[36]

A stunning analysis. Everything is inverted and becomes strictly speaking a game of doubling. The double—which is to say: the ghost of separation— appears, in the final analysis, as the agency that links the real and the psychic or the exterior and the interior. Real or not, trauma is caused by remembered or future separation; it is the cause of *separation that sees itself coming*. The anticipation of mourning for oneself, anticipation as mourning for oneself, the experience of being cut away from the most intimate intimacy at the most intimate point of oneself, is *indestructible*.

This is the central point of our discussion. If, for Freud, the anticipation of danger is the horizon that opens toward every event, nothing seems to threaten this horizon itself. The structure of the effacement of the subject (anticipation of separation) is itself ineffaceable, indelible; it is the indestructible substrate of destruction. In Lacan's words, the horizon of anticipation constituted by the unconscious is at once "fragmented and indestructible."[37] *Never, for Freud, does separation separate from itself.*

"Would you claim," Lacan writes, "that [Freud] solicits [anxiety] only to reduce it to anticipation, preparation, a state of alert, or to a response that is already a mode of defense against what is in the offing? Yes, that is *Erwartung*, the constitution of the hostile as such."[38] Indeed, even though Freud defines trauma as an effraction that penetrates a psyche that does not have time to prepare for this intrusion, he still lays out an entire propaedeutic to this state of unpreparedness which constitutes the originarily anticipatory structure of the psyche. This structure itself cannot be destroyed by the trauma that it anticipates.

Herein lies the crux of the debate between psychoanalysis and neurology with respect to psychic destruction. For the neurologists, the very structure of the cerebral anticipation of death—described as a structure of auto-affection—*is not insulated from danger*, unlike the structure of the unconscious as it is defined by psychoanalysis. Certain types of damage can overwhelm it: the neurological horizon of the anticipation of destruction is destructible.

As a result, certain subjects with brain lesions are deprived precisely of the possibility of seeing or feeling themselves die. A lesion or a synaptic rupture, therefore, can never coincide, symbolically or materially, with anxiety of the cut or of castration.

Lacan and the Thing on the Horizon: Meeting Point and Vanishing Point

Before developing this decisive point, we must further clarify the relation between the unconscious and its destructibility—this time with Lacan, who expands and amplifies the scope of the question.

At first sight, Lacan seems to contradict everything that we have just established about the ineffaceable character of the subject's effacement or the indestructible character of the unconscious structure of the ego's relation to its own disappearance. Indeed, for Lacan, originary separation—characterized as separation of the ego and the subject—is interpreted as the inscription of alterity, or lack, within the ego: "in the circuit, the ego is really separated from the subject by the *petit a*, that is to say by the other."[39]

This separation is the signature of finitude and the form of the structure of anticipating death—a structure that Lacan will call *the horizon of encounter*. But this structure of anticipation is strangely doubled by its *negative*, constituted by a borderline that marks the occurrence of *unforeseeable*, *unthinkable* or *impossible* events—*events that cannot happen*. The set of these "nonevents," in Lacan, are grouped under the name of the Real, "the vanishing point of any reality that might be attained," which is trauma as such, in excess of any horizon of anticipation.[40] This negation of the horizon is indeed the threat that the horizon itself might be destroyed.

The unconscious must then be conceived as the coincidence between *the encounter and the vanishing point*, the conjugated possibility of events that happen and those that are beyond all happening. Trauma, to the extent that it belongs to the second order of events, is thus unassimilable and inappropriable: It resists any transformation into *Erlebnis*, any encounter, any separation anxiety. This is the sense in which, for Lacan, the Real is *beyond* the Symbolic. The Real—or that which is properly traumatic in trauma—does not correspond to any symbol, that is, to *any structure of fragmented unity*, to any "soldering." The Real is inseparable. Without distance, without horizon, without "fissure" or lack: "There is no absence in the real."[41] Lacan declares: "For the real, whatever upheaval we subject it to, is always and in every case in its place; it carries its place stuck to the sole of its shoe, there being nothing that can exile it from it."[42]

Contrary to what we have been arguing, does not psychoanalysis, ultimately, take into account, *besides* events that let themselves be encountered, *other events* that are like holes within the symbolic fabric and exceed any horizon that might encompass them? Freud, as Lacan recalls, distinguishes between two types of events and acknowledges that "*what did not come to light in the symbolic appears in the real.*"[43] Must we not affirm, therefore, that the psychic horizon of anticipation is itself threatened with *real* destruction?

A NEW "SOLDERING": THE THING

In the end, we must answer this question in the negative. It seems, in fact, that a new form of "soldering" intervenes to make it impossible for Lacan, no matter what he says, to think *destruction without remainder.* This new "soldering" would be that of what Lacan calls "the Thing": *das Ding.* This "Thing"—"the absolute other of the subject,"[44] this "prehistoric Other that it is impossible to forget," this "nothing," this "emptiness," this "hole,"[45] this originary trauma—remains, to my mind, that aspect of the Real *that can be said and imagined.* Indeed, the Thing is "that which in the real . . . suffers from the signifier."[46] The Thing is the aspect of the Real that still *happens*; that, in spite of everything, *happens to me.*

The Real knows no lack, it exceeds any horizon of anticipation and, for this reason, can never be encountered. But, for Lacan, *this lack of encounter always converted into a missed encounter.* The Thing is a trope that grants a horizon to this lack of lack and that, in spite of everything, allows the psyche to *see it coming.*

How is it possible to defend such an understanding of the Thing when it seems to run counter to the most widespread interpretations of the concept? A striking chapter from *The Four Fundamental Concepts of Psychoanalysis* is what makes it possible. The title of the chapter, "Tuché and Automaton," speaks for itself. This remarkable text elaborates the concepts of chance and necessity, encounter and real, starting precisely from the signification of the event in psychoanalysis. Lacan undertakes this elaboration through a reading of a dream that Freud discusses in *The Interpretation of Dreams*, the "dream of the burning child."[47]

Lacan recalls that the simplest, least conceptually elaborate meaning of the word *tuché*, which one finds in the work of Aristotle, signifies "fortune," that which happens by chance. *Automaton*, on the other hand, literally designates "that which happens on its own," that which works all by itself and repeats itself according to strict mechanical necessity—as in "automatons" or "automatisms." This distinction can thus account for the two types of event that Freud identifies. *Tuché* could designate the mode in which *Ereignisse* occur—as "pure," unforeseen, and perfectly contingent accidents. *Automaton*, on the contrary, could designate the formation and regulation of the endogenous events that only obey their own law and, in a sense, engender themselves.

Accordingly, Lacan proposes that we translate *tuché* as "encounter with the real." The Real can only occur by chance, *without any machination. Tuché* thus names the specific way in which trauma occurs—since trauma is the Real itself, "unassimilable" by the psyche. To speak of an "encounter with the real," then, amounts to speaking of an impossible encounter. Trauma is what pierces, by force of its contingency, the psychic horizon of anticipation.

The word *automaton* designates "the return, the coming-back, the insistence of the signs by which we see ourselves governed by the pleasure principle."[48] It characterizes the regime of events that, as opposed to the first type of events, can be perfectly well assimilated by the psyche; they are "lived events" that do not overwhelm the energetic circuit, the mechanism, or the automatism of homeostatic regulation.

Has not the distinction between pure accidents (*tuché*) and lived events (*automaton*), events and nonevents, thus been definitively established? Lacan goes so far as to claim that the Real is, in fact, "Freud's true preoccupation as the function of fantasy is revealed to him. . . . He applies himself in a way that can almost be described as anxious, to the question—what is the first encounter, the real, that lies behind the fantasy?"[49] This Real, "there is

no question of confusing [it] with repetition, the return of signs, or reproduction, or the modulation by the act of a sort of acted-out remember-
ing."[50] Lacan insists quite clearly, therefore, upon an essential aspect of the
Freudian thinking of the Real—its status as first trauma without *horizon of
anticipation*.

This is the point at which the concepts become complicated and appear-
ances are inverted. Freud does indeed remark, in *Beyond the Pleasure Principle*,
that trauma tends to repeat itself and that a work of binding (*Bindung*) readies it to be assimilated by the psyche. On the other hand, he finds, as he had
already done in *The Psychopathology of Everyday Life*, that everything associated with *automaton* can also happen by chance. The origin of trauma, as
Lacan recognizes, might only be "apparently accidental"[51]; and the machination of lived or fantasmatic events might only obey the caprice of fortune.
It is impossible to reduce the ambivalence of these terms.

In fact, deeper study of the Aristotelian concepts of *tuché* and *automaton*
reveals that their respective significations are even more ambivalent than
they appear. *Tuché* can certainly be categorized as an exceptional event, but
this does not prevent it from figuring "among things which are for something."[52] On the other hand, *o automatismos* signifies in Greek "that which
happens on its own." But this formula must be understood in two ways:
What happens on its own can do so if it bears its own necessity within itself,
and what happens on its own can also do so, inversely, by pure chance. In
this case, contingency is its self-justification. Aristotle uses *to automaton* precisely to mean "an instance of chance," in opposition to *techné*, technique or
machination. The verb *automatizein* mobilizes both senses at the same time:
to do something by one's own movement or, inversely, to act without reflection, by chance.[53] The chance productions of *tuché*, therefore, all have the
same sense—that is, a certain finality—whereas "automatic" events can
happen by chance. Necessity and contingency thus change places. In the
same manner that the automatism of the event can escape anticipation, the
Real, in the final instance, can be encountered thanks to *the very finality of
its contingency*, can be encountered *as a missed encounter*.

Lacan asks: "Where do we encounter this real? For what we have in the
discovery of psychoanalysis is an encounter, an essential encounter—an
appointment to which we are always called with a real that eludes us."[54] And
this encounter, which is rendered at once possible and impossible by the
tuché, is indeed "essentially the missed encounter."[55]

But is this response satisfying? Does not Lacan himself, at the very moment when he formulates it, miss the hypothesis of *an encounter that would irremediably miss being missed?*

FREUD'S NARRATIVE OF THE "DREAM OF THE BURNING CHILD" AND ITS READING

Let us turn to Freud's narrative of "the dream of the burning child":

> A father had been watching beside his child's sick-bed for days and nights on end. After the child had died, he went into the next room to lie down, but left the door open so that he could see from his bedroom into the room in which his child's body was laid out, with tall candles standing round it. An old man has been engaged to keep watch over it, and sat beside the body murmuring prayers. After a few hours' sleep, the father had a dream *that his child was standing beside his bed, caught him by the arm and whispered to him reproachfully:* "*Father, can't you see I'm burning?*" He woke up, noticed a bright glare of light from the next room, hurried into it and found that the old watchman had dropped off to sleep and that the wrappings and one of the arms of his beloved child's dead body had been burned by a lighted candle that had fallen on him.[56]

Lacan interprets this dream as the dream of an accident but right away raises a troubling question: "Where is the reality in this accident?"[57] *What is real—really accidental—in this accident dream?* Does the accident lie in the candle's falling, the *actual* accident that upsets the sleeper, or is it located *within* the dream—that is, in the sudden appearance of the child who grabs his father by the arm and speaks to him? Is the accident the *tuché* of the fire or the *automaton* of speech, the symbolic manifestation of a reproach? Lacan asks:

> *What is it that wakes the sleeper?* Is it not, *in* the dream, another reality? —the reality that Freud describes thus—*Dass das Kind an seinem Bett steht*, that the child is near his bed, *ihn am Arme fasst*, takes him by the arm and whispers to him reproachfully, *und ihm vorwurfsvoll zuraunt: Vater, siehst du denn nicht*, Father, can't you see, *dass ich verbrenne*, that I am burning?
>
> Is there not more reality in this message than in the noise by which the father also identifies the strange reality of what is happening in the room next door. Is not the missed reality that caused the death of the child expressed in these words?[58]

The reality of the *automaton* is thus more real than that of the flame and the noise; the son's speech is more real than the material reality of falling and burning. The *tuché—flame* and *noise—*is nothing but the means of *another reality* that it allows to happen. Something "more fatal" (the dead son who comes to speak) than (material) reality happens *"by means of* [material] reality."* The *tuché* thus triggers the *automaton*; and the "atrocious vision" of the Thing surges up at the point of convergence between the encounter and the missed encounter: *The Thing occurs as a missed encounter between the father and the son.*

The "material" of the dream, the fortune that gives birth to it, the brute or "idiotic" accident come to pass as such—*ambiguity of the tuché. The finality of the fortuitous accident* consists in the fact that it can only *surrender its place to the symbolic expression of the accident*: "Father, can't you see I'm burning?"

The fact that the symbolic signification of the *automaton* can be inverted to become its mechanical signification only reinforces the former and proves that there is an automatic signification of separation: this sentence, "Father, can't you see I'm burning," is itself a "firebrand—of itself it brings fire where it falls."[59] For the father, it "perpetuates . . . those words forever separated from the dead child that were said to him"[60]—these words cut off from the flesh that they designate, which allows them to be repeated in the absence of the child; these separate words that accomplish separation. The real thus fully projects itself into the field of the symbolic, of the world where we expect it, where we understand it, and where we encounter it.

"What encounter can there be from now on," Lacan asks, "with that forever inert thing—even now being devoured by the flames—if not the encounter that occurs precisely at the moment when, by accident, as if by chance, the flames come to meet him?"[61] Tuché thus reveals that its proper finality lies in the automatism of repetition that, for Lacan, is signifying commemoration: "It is only in the dream that this truly unique encounter can occur. Only a rite, an endlessly repeated act, can commemorate this not very memorable encounter."[62] *The finality of fortune thus lies in symbolic/ automatic repetition—the adventure of separated words.*

Do we not thereby rediscover—no matter what Lacan says—the motif of the indestructibility of the horizon of the encounter, the perspective within which the subject (in this case, the father) is always vulnerable to the event of separation? Nonetheless, to conclude his analysis, Lacan insists upon the indissoluble link between the *repetition compulsion* and the *formation of a horizon*. Reopening the question of the *fort/da* game in *Beyond the Pleasure Principle*, he asserts:

> The ever-open gap introduced by the absence indicated remains the cause of · a centrifugal tracing in which that which falls is not the other qua face in which the subject is projected, but that the cotton-spool linked to itself by the thread that it holds—in which is expressed that which, of itself, detaches itself in this trial, self-mutilation on the basis of which the order of significance will be put in perspective. For the game of the cotton-reel is the subject's answer to what the mother's absence has created on the frontier of this domain—the edge of his cradle—namely, a *ditch*, around which one can only play at jumping.[63]

What appears here, once again, is thus the horizon of anticipation as the horizon of separation from self. The structuring role of castration in the elaboration of this horizon is explicit: "This spool is not the mother reduced to a little ball by some magical game worthy of the Jivaros—it is a small part of the subject that detaches itself from him while still remaining his, still retained."[64] The encounter with the Real as missed encounter does not contradict sexual etiology but rather *confirms* it: "The central bad encounter is at the level of the sexual."[65] If it is true that all events—even "real" or traumatic events—ultimately occur at the heart of the psyche's separation from itself, as proof of this very separability, then the ordeal of cutting, structurally linked to sexuality, remains, for Lacan no less than for Freud, the *indestructible horizon of destruction*.

It is precisely this ineffaceable character of separation—that is to say, of effacement—that, for Lacan, constitutes every trace as a signifier: "The nature of the signifier is precisely that it makes an effort to efface a trace. The more one seeks to efface it, in order to rediscover the trace, the more the trace insists as a signifier."[66] For Freud as for Lacan, separation signifies its own effacement—which is to say that it is never effaced.

Conclusion: The Resistance of Cerebrality

If there remains a chance for cerebrality to resist sexuality, it lies perhaps in the possibility of *abandoning* the paradigm of separation for the idea of a psyche without any structure of openness, a psyche without other on the horizon, without a double. Such a psyche cannot, or can no longer, encounter itself, even by missing itself. Its history has been annihilated. Its trace has finally been effaced along with its becoming-signifier.

Despite Freud's assertions to the contrary, we have seen that there exists a process of cerebral auto-affection. Like any psyche, the brain is constituted by a mirror effect, a reflection of alterity within the loop or "circuit" of selfhood. But this alterity has a particular signification that reveals its full strangeness within the ordeal of trauma. To a certain extent it is a matter of the other *of* the self *in* the self; but, here, *it is the self, and not the other, who never lets itself be encountered* when traumatized. It is the self who is lacking, without specular recuperation.

Structurally, even in its normal state, the cerebral self, or "proto-self," affects itself without encountering itself. As I have said: no one can speak of "his" brain. Between "my" brain and myself there is a sort of opaque wall, an absence of mirror, even as it is the most intimate part of myself, the "me" who thinks and feels within "me."

The activity revealed by brain imaging—the only mirror that can objectify cerebral auto-affection—is what makes narcissism possible to the extent that such photographs of *my* brain are necessarily offered, without any possible internalization, to *the gaze of the other*, even if this gaze is *my own.* However, the eye of the other upon my most living intimacy, the eye of the other upon my thinking and feeling connections, the eye of the other upon the meshwork of my affects, which affect themselves and affect me without gazing at themselves or me gazing at myself, the eye of the other does not give birth to any mirror stage. *Cerebrality does not gaze at itself.*

On the cerebral level, there is not really any "reception of the self in the other"—if one thereby understands the movement that Lacan calls the "Imaginary." There is no line of sight, no constitution of subjectivity in "seeing oneself in the gaze of the other," no struggle for recognition. The regulation and organization of the brain, in fact, cannot be accounted for either in terms of the "Real" or the "Imaginary."

We take the risk of introducing a fourth instance into the program of Lacan's Real-Symbolic-Imaginary: the "Material." The Material would constitute the sense of an affective economy that solicits itself *without seeing itself.*

If the "normal" brain never affects itself without meeting up with itself in the mirage of its own separation, the damaged brain has no chance, *a fortiori*, of being present to its own fragmentation or its own wounding.

Contrary to castration, no representation, no phenomenon, no example of separation can make it possible to anticipate, foresee, or fantasize the aftermath of a ruptured cerebral connection. Such a transformation cannot even be dreamed. There is no scene for *this* Thing that is not the Thing. The brain in no way anticipates the possibility of its own damage. When damage occurs, it is another self who is affected, a new self, unrecognizable.

The resistance of cerebrality to sexuality, in the final instance, pertains to the manner in which the cerebral self *belongs to the other without alienation or specularity*.

What scorches the symbolic is the *material* destruction of the Thing.

EIGHT

Neurological Objection: Rehabilitating the Event

Psychotherapeutic experience on the battlefield gave birth to a renewal of the paradigm of psychoanalysis itself.

— FRANÇOISE DAVOINE AND JEAN-MAX GAUDILLIÈRE, *History Beyond Trauma*

All those who "saw themselves dead" are psychically traumatized. It is not only a matter of the "imaginative" apprehension of death, but the sudden and inaugural perception of "one's own death" as something imminent and unavoidable: the revelation of something mysterious, grave, and definitive that has no meaning except perhaps as an introduction into nothingness.

— LOUIS CROCQ, *Les traumatismes psychiques de guerre*

Psychological trauma is an affliction of the powerless. At the moment of trauma, the victim is rendered helpless by overwhelming force. When the force is that of nature, we speak of disasters. When the force is that of other human beings, we speak of atrocities. Traumatic events overwhelm the ordinary systems of care that give people a sense of control, connection, and meaning.

— JUDITH LEWIS HERMAN, *Trauma and Recovery*

At the end of August 1914, a Lieutenant Kauders was wounded near Lublin by Russian gunfire. On September 9, nine days after being wounded, he was diagnosed with a skull fracture. Kauders had trouble walking, the two sides of his body fell out of unison, his vision was impaired, and he suffered from affective disturbances and apathy. The Arbitration Commission declared him an invalid and sent him home to Berlin.

Later, in 1917, probably because officers were hard to come by, his classification as an "invalid" was reevaluated. That autumn, Kauders appeared once again before the military commission, which ordered him to report to Vienna. To his great astonishment, he was sent to Garrison Hospital No. 1 and placed in a room like a prison cell with iron bars. His neighbor, a Turkish officer who was clearly mentally ill, screamed all through the night. The following day, Kauders was taken by ambulance to the Wagner-Jauregg clinic, where he remained in an isolation room for seventy-seven days. The deplorable hygienic conditions and the brutality of the guards were nothing in comparison to the sessions of faradization, which Wagner-Jauregg implemented himself. A metallic brush charged with electric current was applied to sensitive parts of his body such as the testicles and the tips of the toes while an audience shouted at him: "Miserable malingerer!" "You're in for it!" The pain was intolerable: "On treatment with a strong electric current, the patient begins to weep loudly and lament with pain."[1] "I was quite intentionally exposed to torture," Kauders claims in his report.[2]

At no moment did the psychiatrists accept the diagnosis of brain damage, and so they focused their efforts on using electricity to make the patient confess to malingering. Unable to extract such a confession, they ended up discharging Kauders with the diagnosis of "traumatic hysteria." Eissler comments: "I have referred to all the medical findings and reports. . . . Their gist is this: The greatest German neurologist and an important Viennese specialist in internal medicine were sure, or nearly sure, that Kauders had suffered an organic injury to the cerebrum. Wagner-Jauregg was sure that Kauders was a malingerer."[3]

The neurological diagnosis was thus fallaciously transformed into a psychiatric diagnosis. Nonetheless, detailed examination of the expert neurological reports shows that, if Kauders was not the victim of serious damage to the central nervous system, he did suffer from a hematoma that formed after a skull fracture. Eissler asserts:

> It was already known at that time that the absence of external injury by no means excluded internal damage. As early as 1895 Strümpell expressed himself unambiguously about the matter: "Even after relatively slighter wounds (caused by thrusts, blows and the like) we can rarely dispute the possibility of resulting hemorrhages, either in the meninges and the surface of the brain or in the deeper part of the brain" . . . Sänger wrote (in 1915): "After grenade explosions, even where there is no external injury, peculiar disturbances occur, the cause of which needs further investigation (meningeal hemorrhage and the like" . . . Bonhoffer (1917) wrote: "The anatomical findings of multiple clinical

hemorrhages in the brain, which we meet in brain-tissue specimens after grenade explosions without external injury . . . leaves no room for doubt that this type of organic damage (by air pressure in grenade explosions) can occur."⁴

One wonders, then, why neither Wagner-Jauregg, the military doctor, nor Freud, during his expert testimony, did not recognize the internal wound and did not want to admit that Kauder's troubles, in particular the loss of feeling in his body, resulted from a subdural hematoma. Neither the diagnosis of malingering nor that of a withdrawal of libidinal cathexis after a shock could account for Kauder's physical and *psychic* suffering. Eissler has already said it: *only a dual approach, both neurological and psychoanalytic—a prefiguration of neuropsychoanalysis—could help this man.*

How could Freud remain so blind to this point? And what theoretical or clinical conclusions should we draw from the contemporary neurological reexamination of traumatic neuroses in general and war neuroses in particular?

Freud's "Abandonment" of the Concept of Traumatic Neurosis

As his thought develops, Freud abandons both the theory of *the traumatic origin of neuroses* and the theory of specific psychic disturbances that follow from *traumatic neuroses*. In the same way that the role of "accidental influences" become relativized in the texts devoted to the formation of infantile sexual life, the causal power of traumatic events, such as wars or accidents, finds itself systematically subordinated to the more originary legality of the "libido theory." The more that Freud takes such events into account and examines them in their own right, it would seem, the less he endows them with etiological value.

Nonetheless, the article entitled "Traumatic Neurosis," from *The Language of Psychoanalysis*, concludes:

> It may thus be seen how psychoanalytic investigation throws the concept of traumatic neurosis into question: it contests the decisive function of the traumatic event—first by stressing its relativity vis-à-vis the subject's tolerance, and secondly by inserting the traumatic experience into the context of the subject's particular history and organization. . . . The notion of traumatic neurosis appears as nothing more than an initial, purely descriptive approximation which cannot survive any deeper analysis of the factors in question.⁵

It would even be possible to say that, for Freud, traumatic neurosis, in the strict sense of a psychic disorder caused by piercing or effraction, does not exist.

In this regard, an assertion from *Beyond the Pleasure Principle* is emblematic: "At this point I propose to abandon the dark and dismal theme [*dunkle-dünstere Thema*] of the traumatic neurosis."[6] In "Introduction to *Psychoanalysis and the War Neuroses*," Freud already claimed that war neuroses were destined to disappear with the armed conflicts that caused them. The Fifth Psychoanalytic Congress concluded with the resolution to establish psychoanalytic centers "at which analytically trained physicians would have leisure and opportunity for studying the nature of these puzzling disorders and the therapeutic effect exercised on them by psychoanalysis." But, Freud continues,

> Before these proposals could be put into effect, the war came to an end,
> the state organizations collapsed and interest in the war neuroses gave
> place to other concerns. It is, however, a significant fact that, when war
> conditions caused to operate, the greater number of neurotic disturbances
> brought about by the war simultaneously vanished. The opportunity for a
> thorough investigation of these affections was thus unluckily lost—though,
> we must add, the early recurrence of such an opportunity is not a thing to be
> desired.[7]

It is thus evident that Freud abandons the problem. From his early texts on hysteria to *Beyond the Pleasure Principle* or *Civilization and Its Discontents*, by way of the texts specifically devoted to anxiety, it seems that Freud is less concerned with clarifying his thinking of the event than with the sexual etiology of the event. The latter becomes more and more differentiated, to the point that its causal structure develops two heads united in a single theory—the "libido theory." This structure, both bifid and unified, is then imposed, between life and death, as an incontestable principle of explanation.

Freud's position on the question of the accident remains, from one end of his work to the other, fundamentally the same. These lines from *New Introductory Lectures on Psychoanalysis* encapsulate this position: "We ask ourselves what it is that is actually feared in a situation of danger of this kind. It is plainly not the injury to the subject as judged objectively, for this need be of no significance psychologically, but something brought about by it in the mind."[8] The problem is always that of the internal enemy or the enemy interior. But the question of a proper body that becomes an enemy through the effect of an improper foreign body—bullet, shell, explosion—is never truly envisaged.

Accordingly, Eric Porge's analyses, in his remarkable introduction to Eissler's book, remain quite debatable, for they claim that *Beyond the Pleasure Principle* offers a radically new theory of trauma that will finally push Freud to break with the doctrines of his students: "All the efforts of Freud's students were devoted to showing that the same principles were at work in the war neuroses and in the neuroses of peacetime: it was important that war neuroses should confirm the doctrine of the conflict between the ego and the libido."⁹ Freud would thus literally invert this perspective. Porge continues: "Certainly, in writing that the war neuroses are analogous to the transference neuroses . . . Freud authorizes an approach that begins with the old principles."¹⁰ But, in *Beyond the Pleasure Principle*, "far from retracting the word trauma . . . Freud actually returns to it and lends it a new dimension."¹¹ The inversion is thus formulated in the following terms: Freud "abandons" the theory of sexual character in order to affirm the traumatic character of the sexual: "At the end of a long trajectory, Freud once again comes up against the traumatic character of the sexual. Such is the inversion that he accomplishes. Recognizing trauma as a 'source of sexual excitation,' Freud turns toward a sexual origin other than that of the 'conflict between the ego and the sexual drives that it represses.'"¹²

This analysis is not false—except that, in its own terms, what Freud redefines is sexuality, not trauma. The two characteristics of sexuality that Porge quite rightly distinguishes participate not in a redefinition of trauma but rather in the redefinition of instinctual dualism, or the libido theory. *The sexual is what changes sense, not trauma. The structure of the event, traumatic or not, remains the same.*

New Baptisms: Posttraumatic Stress

War psychiatry has also abandoned the categories of "traumatic neurosis" and "war neurosis." But the reasons for this abandonment are obviously not at all the same as Freud's. The extreme gravity of psychic disturbances caused by war meant that it would no longer be pertinent to characterize them in terms of "neurosis." As opposed to what Freud claims, the traumatic illnesses occasioned by war do not disappear after the war is over.

On the contrary, they become so much more complex and severe that the concept of neurosis can no longer account for them. Indeed, as the military psychiatrist Louis Crocq has remarked in *Les traumatismes psychiques de guerre*, psychiatry has successfully called into question the subordination of wound and trauma to narcissism—that is, the subordination of the lesion to certain types of libidinal investment. The screen that separates the organic wound and the constitution of a neurosis or psychic disturbance has been lifted.

We now know, Crocq asserts, that Freud's claim that wounds prevent the onset of neurosis is not true. Certainly, he continues, "the fact of having a somatic wound does not necessarily cause a traumatic neurosis," but "it does not protect against its onset."[13] It also untrue, therefore, that war neuroses disappeared after the end of the First World War. There have obviously been further wars and thus further war neuroses; but, more important, psychopathology has become increasingly preoccupied with the wounds that these wars left behind. This is so much the case, Crocq writes, that "the question arises once again whether it might in fact be trauma that brings about obsessions, rituals, and compulsions of destiny, or whether these characteristics are the traits of a subject who is obsessional beforehand and compromised in advance by intrusive thoughts and behavior."[14]

THE WORK OF ABRAM KARDINER

If the concept of "traumatic war neurosis" has fallen out of use, it is because it became necessary for military psychiatrists to propose a new clinical category that would do justice to *brutal shock*, to the *impact of the event* in the psychic constitution of trauma. It was the American psychoanalyst Abram Kardiner, a former patient of Freud, who was the first to put war psychiatry on the path toward the decisive changes that led to abandonment of the concept of traumatic neurosis in favor of *posttraumatic stress disorder*.

When he returned to New York after his Viennese psychoanalysis, Kardiner worked both in his own analytic practice and at the veterans' clinic, where he studied numerous cases of war neurosis. At first, he attempted to develop a new theory of the war neuroses within the framework of classic psychoanalysis; but he abandoned these efforts and, out of disappointment, turned for a time to anthropology.[15] It was only in 1941 that he returned to the problem of the war neuroses and established his new thinking in his fundamental work *The Traumatic Neuroses of War*, revised in 1947 in collaboration with Herbert Spiegel.[16]

Kardiner, who first proposed the concept of "traumatic neuroses of war," developed the idea that war traumas impact the "effective ego" rather than the "affective ego" of the soldier.[17] The mechanisms of defense at work in "peacetime" neuroses are not required, as Freud had asserted, by disturbances linked to traumatic neuroses because the latter derive from *emergency* measures on the part of the "effective" ego. The "effective ego" designates the secret part of the psychic organization charged with eliminating aggressive stimuli through the alteration of adaptation. Its principal functions are "perception (including meaning and use, voluntary motion, orientation, memory, inhibition, and repression)."[18] Traumatic experience occurs when the effective ego finds itself out of action. Under such conditions, its only recourse is to repress itself. This results in the phenomenon of "constriction" that manifests itself as *affective withdrawal*.

Kardiner developed the hypothesis of an "overwhelming" of the psyche that manifests itself through four decisive neuropsychological symptoms: "Changes in the perception of the external world, changes in the technics of adaptation, changes in self-perception, and changes in neurovegetative life."[19] It is the concept of suddenness that emerges within the analysis of these disorders and that determines the specificity of traumatic experience. Accordingly, a *therapy of emergency* must, from then on, respond to the *unforeseen* character of the trauma.

THOMAS SALMON AND THE PSYCHIATRY OF EMERGENCY

The American psychiatrist Thomas Salmon, sent to Europe on a mission to observe the phenomena of the war neuroses before the Americans entered the conflict on April 2, 1917, already developed the idea of a therapy of *immediacy*, now known under the name of *forward psychiatry*. "Forward" should be understood to mean "up at the front." Salmon noted, in fact, "that soldiers treated near the battlefront, close enough to hear the sound of cannon fire, recovered more quickly than if they had been treated in the rear."[20] In his report, entitled *The Care and Treatment of Mental Diseases and War Neuroses (Shell Shock) in the British Army*, published in New York in 1917, Salmon insists on the dangers linked to length of time it took to evacuate and repatriate traumatized soldiers because such a delay would give these soldiers time to consolidate their morbid mentality.[21]

Salmon developed the idea of creating hospitals close to the front where doctors would work shoulder to shoulder, which would constitute true emergency therapeutic support. These hospitals prefigure the present-day emergency medical cells that are set up in disaster zones (terrorist attacks, explosions, fires, accidents), where they use *debriefing* techniques to treat traumatic experience.[22]

Salmon articulated five principles of emergency therapy: Proximity, Immediacy, Expectancy, Simplicity, and Centrality. Proximity entails keeping the soldier for a time within the atmosphere of the front. Immediacy responds to the need to prevent the development of a traumatic neurosis. Expectancy implies that the patient must be convinced that he will get better. Simplicity entails rejecting the use of over-complicated equipment or technical apparatuses: *simple* offices, clean but rustic, must suffice; and the treatment itself, founded upon "persuasive suggestion," must also be simple. Finally, centrality suggests that the medical organization must be centralized, fanning outward from the foremost centers to the rearmost annexes, annexes where the most serious cases, often patients without hope of survival, would be transferred.

THE POSTTRAUMATIC STRESS DISORDER DIAGNOSIS AND ITS GRADUAL ENLARGEMENT

In civilian life, it took a long time for specific symptoms linked to the traumatic experience of war to gain widespread recognition. Many soldiers complained that their suffering was not recognized when they returned from war. Accordingly, it was only in 1980 that that concept of posttraumatic stress disorder (PTSD) appeared for the first time, when it figured in the DSM-III as a category designed primarily to characterize the mental state of Vietnam War veterans. It was thus long after the war itself that the symptoms caused by the war were identified and recognized.

The clinical profile of PTSD comprises: (1) exposure to a stress-inducing event that causes distress in a subject; (2) reminiscences of the event (dreams or obsessive memories); (3) psychic exhaustion manifesting itself as loss of interest in habitual activities, a tendency to detach oneself from others, and a constriction of affect (coolness or indifference); (4) a disparate set of symptoms, such as loss of memory, state of alert, problems sleeping, survivor's guilt, and the aggravation of these symptoms when the patient is exposed to stimuli that recall the trauma. Ruth Leys writes:

> Post-traumatic stress disorder is fundamentally a disorder of memory. The idea is that, owing to the emotions of terror and surprise caused by certain events, the mind is split or dissociated; it is unable to register the wound to the psyche because the ordinary mechanisms of awareness and cognition are destroyed.[23]

It quickly became apparent that this clinical picture did not only apply to disorders linked to war, but that it could also easily pertain to *any type of trauma*. It thus became necessary to extend PTSD to cases beyond the old cases of "war neuroses," even if, once again, official recognition of this necessity took time. Louis Crocq writes, "As such clinical findings mounted and as clinicians gained more and more experience with PTSD in accident or disaster victims, the DSM commission twice modified the diagnostic criteria for PTSD: first in 1987, in the DSM-III-R, and then in 1994, in the DSM-IV."[24] Accordingly, "in the most recent version (from 1994), the criteria of the traumatic event (criteria A) now requires that the event be experienced in the mode of fear, impotence, and horror; and the criteria of reminiscence (criteria B) now requires distress."[25] It thus took fourteen years, between 1980 and 1994, to recognize the link between the traumatic factor and the experience of fear, impotence, or horror provoked by an *objective event* that may not have to do with war.

The threat of death, wounding, suffering, torture, violence, the spectacle of an other's death or suffering, the sight of material destruction (houses, villages) are now officially part of "traumatic etiologies."[26] In her book, *Trauma and Recovery*, Judith Lewis Herman asserts, "A psychological trauma is an affliction of the powerless. . . . Traumatic reactions occur when action is of no avail. When neither resistance nor escape is possible, the human system of self-defense becomes overwhelmed and disorganized."[27] In the history of the PTSD diagnosis, therefore, it is possible to measure *the profound displacement of the concept of traumatic neurosis as psychoanalysis had defined it*. Not only does this diagnostic criteria make it possible more precisely to identify the psychic disturbances of war, it also calls into question, if not *dismantles, the very link between neurosis and trauma*. The impotence linked to traumatic shock can no longer be returned to the native "helplessness" of the subject. No "internal conflict" can be invoked. The "soldering" that joins *Erlebnis* and *Ereignis* is no longer possible. The psychic past no longer functions as resource for the present. The past is no longer the reason for trauma. PTSD allows for the existence of events that, unto themselves, are *their own origin*, and that, by virtue of their specific power, occasion a new psychic life. These events are precisely traumatic events.

Rehabilitating the Event

THE NEUROTIC IMPOSTURE

From one version of posttraumatic stress disorder to another—both the diagnosis in military contexts and its enlarged application to situations of grave danger in civilian life—what is at stake is *a radical redefinition of traumatic etiology*. This redefinition revolves around the imperative, which Louis Crocq formulates, to "rehabilitate the event": "In matters of traumatic psychopathology, we must guard against the temptation to fall back upon predisposition, or even predestination. . . . It is time to rehabilitate the event—its destructive, overwhelming, and disorganizing power."[28] This "rehabilitation of the event" transforms the Freudian conception of the "soldering" event. Certainly, a traumatic accident can indeed reawaken or reactivate an "old event" and thereby an internal conflict.[29] But today psychiatrists and neuropsychiatrists insist upon the unexpected and irreducible character of the traumatic event, which, even if it recalls past trauma, cannot do so without *profoundly modifying the vision and content of the past itself*.

By virtue of its pathological force of deformation and its destructive plasticity, in fact, such an event introduces an *inauthenticity*, a *facticity* within psychic life. It creates another history, a past that does not exist and, in this sense, constitutes a "neurotic imposture."[30] It is not only the behavior of the hysteric that mimes cerebral pathology; cerebral pathology itself is also capable of *miming neurosis*:

> The notion of neurotic imposture, or the invasion of the personality by an overwhelming event, and then by its pseudo-memory that has become foreign and exclusionary, is well suited from a phenomenological perspective to account for the experiences of foreignness, depersonalization, dispossession, and xenopathy—that is, of alienation—to which patients have testified. . . . It is an entire experience of constricted and inauthentic being-in-the-world, with its

infiltration into the present, its obstruction of the future, and even its reorganization of the past, all of which constitute the unfortunate destiny of the traumatized neurotic.[31]

TRAUMATIC METAMORPHOSIS

The specificity of the traumatic event thus inheres in its *metamorphic power*. The traumatic event, in a certain sense, invents its subject. The past of the traumatized individual changes, becomes *another past* when it is not pure and simply destroyed or consigned to oblivion. Accordingly, a *new subject* enters the scene in order to assume this past that never took place. It is no longer the same subject who anticipates himself and sees himself die. Separation can no longer be anticipated but it does occur, precisely, in metamorphosis.

Crocq insists on the "lasting modification of the personality that follows an experience of catastrophe," a modification that leads him to propose the concept of the "traumato-neurotic personality."[32] Further: "The personalities of today's neuroses are 'constitutional' . . . *the traumato-neurotic personality is only constituted through trauma, at any age. It is what becomes of the personality of the patient under the impact of trauma.*"[33]

This modification of the identity of patients is a long-established fact of observation. Simmel already spoke of a "change in the soul" and an "entombment of the person" under the effect of the accident.[34] But the psychoanalytic model, "even as it accounts for the effraction and its overwhelming effect on the personality, *does not postulate that this personality has been changed.*"[35] If Freud admits the obvious fact that an accident can severely and permanently damage the psyche, he never presents the formation of the new identity as a discontinuous process, a leap, a phenomenon that is no less unforeseen or unexpected than the catastrophe itself. For him, traumas and wounds do not seem capable of creating *ex nihilo* a posttraumatic identity. There is always a certain psychic continuity between what comes before and what comes after the wound; the subject remains what he is within his very alienation. However, this very continuity is what will be called into question within contemporary neurological debate.

Indeed, from the neurological point of view, the hypothesis of absolute danger designates the risk of brutal and sudden disappearance of the trace resulting in the formation of an identity without origin and without memory;

an identity, produced by destructive plasticity, that is not interested or only falsely interested in itself.

Rehabilitating the event is thus a matter of taking into account the discontinuity produced by the traumatizing event and of its destructive power to transform identity ("Gage was no longer Gage").

> If we ask patients about their experiences of these changes of personality, we observe that this is no metaphor. The patients find themselves really changed; they no longer recognize themselves as they were before. And this is not simply due to the fact that they are sad about having undergone a difficult event; it is, more profoundly, on the level of their entire way of living, that they come to realize that a new being is within them, a being whom they do not recognize.[36]

THE SUBORDINATION OF SEXUAL ETIOLOGY

The contemporary conception of the traumatic event has permanently disrupted the link between mechanical commotion and sexual excitation that was first established by Freud and subsequently consolidated by Fenichel. In his work, *The Psychoanalytic Theory of the Neuroses*, Fenichel accords much importance to the genital excitation that subtends traumatic excitation. Many traumatized people, in fact, fear their own sexual excitation: "Some types of morbid fear of death when analyzed turn out to be fears of orgasm"[37] Anxiety would emerge from a displacement of excitation from the genital domain to the neurovegetal domain. Fenichel even goes so far as to affirm that certain "traumatophilic" subjects "appear to seek out the traumatic situation unconsciously—although they dread it at the same time."[38] Such assertions lose their relevance when they come up against the clinical problems posed by subjects suffering from PTSD, which all revolve around a brutal intrusion from outside that escapes any anticipation and any encounter with an "inside" prepared to host the brutality of the effraction.

Sexuality loses its etiological value as a primary cause to the extent that the regime of events that it governs cannot or can no longer integrate the traumatic event—henceforth defined as "something that occurs one time, at a specific date and time, within a person's history or that of a human society, the signification and consequences of which impinge upon everything else that has happened (and, sometimes, that will happen) during the rest of this history."[39] The DSM-III specifies that the trauma is a "psychologically distressing event that is outside the range of usual human experience," going beyond merely painful or unpleasurable experiences (such as "simple bereavement, chronic illness, business losses, and marital conflict").[40] This "stressor" represents a serious threat to the life and physical integrity of the subject. It introduces a radical cut between the past and the present.

The impossibility of confusing traumatic factors and psychoneurotic factors has thus been established, which deprives sexuality of its etiological validity.

Toward a General Theory of Trauma

This redefinition of trauma transforms the entire field of psychopathology, not merely the field of war psychiatry.

In his commentary on the stunning film, *L'effroi des hommes* (Human Dread), directed by Jean-Bernard Andro, the military psychiatrist Claude Barrois opposes the *Orpheus myth* to the *Oedipus myth*. Traumato-neurotic personalities have returned from hell not from childhood.[41] "This film brings together the testimony of survivors in a chiasmus: the memory of hell and the hell of memory. . . . Back from the land of the dead, the poet is no longer inclined to sing."[42] Barrois shows that everything that war psychiatry has revealed about the subject of trauma not only applies to the victims of armed conflicts but also *all subjects whose identity has been profoundly metamorphosed from the impact of a catastrophe.*

> Whatever their age, origin, profession, sex, or condition; whatever the circumstances of the trauma (attack, airplane hijacking, wars, accidents, natural disasters); whatever the degrees of their exposure; with or without injuries; alone or lost in a crowd; on the domestic, national, or international scale, in their accounts these survivors sketch out a field of research that their symptoms are exploring.[43]

The substitution of Orpheus in place of Oedipus, the substitution of a hell without memory in place of a purgatory of culpability pertains to all the new wounded whether or not they have returned from the field of battle.

Judith Lewis Herman also underscores that today's wars have manifestly blurred the distinction between military and civilian.[44] The concept of war

trauma has thus been extended to civilian victims. The psychic consequences of violence against women and children during military conflicts,
the symptoms of captivity, the disorders suffered by hostages or victims of
terrorism, and the stress experienced by peacekeepers or humanitarian per-
sonnel can all be understood in terms of this concept. It would even be
necessary, in Herman's view, to enlarge the scope of posttraumatic stress
disorder by renaming it complex posttraumatic stress disorder in order to
designate a multiple and differentiated—one might even say, *universal*—
state of stress.[45]

The concept of PTSD should thus extend to *any and all cases of trauma.*
As I affirmed at the beginning of this study, the catastrophic event is itself
void of sense and traumatic experience is first and always an experience of
the absence of sense.[46] Moreover, it is striking to note that today's victims of
sociopolitical traumas present the same profile as victims of natural catastrophes (tsunamis, earthquakes, floods) or grave accidents (serious domestic
accidents, explosions, fires). We have entered a new age of political violence
in which politics is defined by the renunciation of any hope of endowing
violence with a political sense.

The meaning of armed conflicts, for example, is masked behind the
impersonal and signatureless character of their attacks. Between a car bomb
and an accidental detonation of a gas tank there is both an enormous difference and no difference. The sinister lesson of terrorism lies in its refusal to
formulate a lesson. Responsibility for attacks is claimed less and less often.
The situation in Iraq, for example, remains illegible: Who perpetrates terrorist attacks today, and why? The dissimulation of the reason for the event
is the new form of the event. The increasingly radical effacement of the
distinction between accident and crime, between disastrous incidents and
war, the multiform presence of the absence of any responsible instance
or author makes *the natural catastrophe of contemporary politics* into a daily
occurrence.

The sheer number of these traumatic events tends to neutralize their
intention, such that they assume the unmotivated character of the chance,
uninterpretable event. *The enemy, today, is hermeneutics.* This is why it falls
to neurology, psychoanalysis, and neuropsychoanalysis, starting from the
redefinition of trauma, *to produce the sense of this war on sense.*

This effacement of sense, of course, does not only occur in countries
at war; it is *everywhere.* It constitutes the new face of the social—bearing

witness to an emergent, *globalized* psychic pathology that is identical in all cases and all contexts. "Psychological trauma," Judith Lewis Herman declares, "is indeed a worldwide phenomenon."[47]

It is perhaps surprising that all these traumatic events should be placed on the same level. I might be reproached for forging a false "amalgam," or for following the method that Boris Cyrulnik elaborates in his book *Un merveilleux malheur* (A Marvelous Misfortune), where he applies the concept of trauma to victims of the Nazi concentration camps, Romanian orphans, abused children, children caught up in the horrors of the war in Mozambique, or famous accident victims. But what must be discussed today is precisely this phenomenon of the amalgam, *the heterogeneous mixture of nature and politics* at work in all types of violence, this mixture where politics is annulled as such so that it assumes the face of nature and where nature disappears beneath the mask of politics. *This globalized heterogeneous mixture of nature and politics is brought to light by the worldwide uniformity of neuropsychological reactions.*

The Etiological Triumph of Cerebrality

THE NEW WOUNDED AND THE "PARADIGMATIC" STATUS OF BRAIN DAMAGE

In the course of this analysis, I have attempted to make my way back to people with brain lesions and those traumatized by war; to people traumatized by war and war's civilian victims; and, then, to *all* traumatized people. The new wounded come together around a single fact: the radical rupture that trauma introduces into the psyche.

This rupture coincides with the theoretical trauma that rattles the doctrinal body of psychoanalysis. The shattered psyches of the new wounded fissure this body with a new eventality—the caesura of new axiological regime of events that is none other than cerebrality.

In the course of my argument, I have granted brain lesions the status of paradigm—in the two senses of this term. The paradigm is the most exceptional, the "most exemplary" of all examples and at the same time the most banal of all examples; it is both *the* example and *an* example among others. Brain lesions are paradigmatic in the first sense in that they make it possible to construct a descriptive model for all posttraumatic behavior. They are paradigmatic in the second sense in that they only represent one case among others of this model itself.

Brain lesions are paradigmatic in the sense that they are *the very example* of violent, meaningless, unexpected, and unforeseeable shock that transforms the identity of the subject, interrupts his relation to himself and permanently disorganizes the process of his auto-affection. At the same time, however, they are *nothing but examples* of this model, traumas *among other traumas*, no different than the others, phenomena that share the same characteristic traits with the others.

Cerebrality can constitute both a law and a case of its own law because all traumas—not exclusively brain lesions—are accompanied by brain damage. As proof of this claim, we need only point to the fact that they all entail indifference and affective coolness or attacks upon the emotional brain. Judith Lewis Herman writes:

> The features of post-traumatic stress that become most exaggerated in chronically traumatized people are avoidance or constriction. When the victim has been reduced to a goal of simple survival, psychological constriction becomes an essential form of adaptation. This narrowing applies to every aspect of life—to relationships, activities, thoughts, memories, emotions, and even sensations. And while this constriction is adaptive in captivity, it also leads to a kind of atrophy in the psychological capacities that have been suppressed and to the overdevelopment of a solitary life.[48]

In every case, "indifference, emotional detachment, and profound passivity"[49] figure among the symptoms attached to the state of posttraumatic stress. *Affective barrenness* is the trait that all these states have in common: loss of curiosity, loss of motivation, disinterest in close friends and relatives, withdrawn behavior.

There is no trauma, therefore, without impact upon cerebrality. Cyrulnik asserts: "In the wake of aggression, *metamorphosis is biological*."[50] This metamorphosis, born of the wound, deeply transforms "ways of seeing and responding to the world."[51] Alan Shore has extensively analyzed the effect of childhood traumas linked to abuse upon the frontolimbic region of the brain:

> Social-emotional environments that provide traumatizing attachment histories retard the experience-dependent development of the frontolimbic regions, especially the right cortical areas that are prospectively involved in affect

regulating functions. . . . In the last decade, a growing body of neurobiological research on PTSD has uncovered dysfunctional frontal-subcortical systems and an altered functional activity of the orbitofrontal cortex, anterior cingulate, and amygdala.[52]

The perturbation of affects in posttraumatic states extends well beyond cases of brain lesions and thereby assumes universal relevance.[53] Biological testimony to the fact of cerebrality, or cerebral causality, far from "reducing" suffering to objective data, *reveals another sense of the psychic event that depends upon its absence of sense.*

ANOTHER LOOK AT THE LINK BETWEEN BIOLOGICAL VIOLENCE AND POLITICAL VIOLENCE

This is to say that neuropathology today reopens the great question of the relation between biology and the social. The objective neurological impact of trauma makes it possible to sketch a new worldwide typology of psychic illness that, pertaining neither to neurosis or psychosis, allows the disaffected faces of the victims to appear at the border between nature and community.

As we look at these faces, it is impossible to forget what unites them and effaces the distinction between lesional trauma, sociopolitical trauma, and trauma caused by natural cataclysms. The differences among the sources of such wounds tend, in fact, to become blurred on the level of their effects (phenomena of emotional barrenness and withdrawal).

It is notable that neurologists never present cases of brain lesions without placing them in a social context. Moreover, they invite us to treat such cases themselves as *political cases.* In the opening pages of his book *The Feeling of What Happens*, Antonio Damasio meditates on the figure of an old man walking through the streets of Stockholm. From a window, he watches "a frail old man make his way toward a ferry that is about to depart. Time is short, but his gait is slow; his steps break at the ankle from arthritic pain; his hair is white; his coat is worn . . . his whole body seemingly saying, Is this it? Am I in the right place? Where to next?"[54]

The vision of this lost old man immediately awakens a memory:

Thirty-two years ago, a man sat across from me in a strange, entirely circular, gray-painted examining room. The afternoon sun was shining on us through a skylight as we talked quietly. Suddenly the man stopped, in midsentence, and his face lost animation; his mouth froze, still open, and his eyes became vacuously fixed on some point on the wall behind me. For a few seconds he remained motionless. I spoke his name but there was no reply. Then he began to move a little, he smacked his lips, his eyes shifted to the table between us, he seemed to see a cup of coffee and a small metal vase of flowers; he must have, because he picked up the cup and drank from it. I spoke to him again and again and he did not reply. He touched the vase. I asked him what was going on, and he did not reply, his face had no expression. He did not look at me. . . . I got up and called him again. He stopped, he looked at me, and some expression returned to his face—he looked perplexed. I called him again, and he said, "What?"

For a brief period, which seemed like ages, this man suffered from an impairment of consciousness. Neurologically speaking, he had an absence seizure followed by an absence automatism, two among the many manifestations of epilepsy, a condition caused by brain dysfunction.[55]

How could we not be struck by the obvious similarity between the general comportment and behavior of a social outcast and a person with a brain lesion? How could we avoid drawing a connection between neuropathological disaffection and "disaffiliation"?[56]

The first reason for making this connection is certainly the fact that people with brain lesions are often liable to be excluded from the social bond because of their handicap. Their emotional coolness is doubled by an inability to choose, to accord weight to things, to make good decisions and such deficits often make it impossible for the subject to integrate himself into the work world. Just as reason comes up short when deprived of emotional signals, the will cannot will anything in the absence of affect. People with brain lesions most often suffer from a collapse of social status because of their inability to decide, to choose, and to will.

This is why it is possible to consider neuropathologies as "sociopathies." The case of Phineas Gage furnishes a good example of this link between biology and the social system:

When the neuronal machinery that specifically supports the buildup and deployment of somatic markers is damaged in adulthood, as it was in Gage, the somatic-marker device no longer functions properly even if it has been normal until then. I used the term "acquired" sociopathy, as qualified shorthand, to describe a part of the behavior of such patients.[57]

The second reason for the relation between biology and the social derives from the fact that the social itself can be the cause of traumas that induce behaviors analogous to those of neuropaths. Certain acts of extreme violence appear to be dictated by a certain behavorial absence, by the dysfunction of emotional markers—in other words, by a disturbance of cerebrality engendered by the sociopolitical context.

> There are many Gages around us, people whose fall from social grace is disturbingly similar. Some have brain damage consequent to brain tumors, or head injury, or other neurological disease. Yet some have had no overtly neurological disease and they still behave like Gage, for reasons having to do with their brains or with the society into which they were born.[58]

Such analyses lead us to take into account the "sociopolitical faultlines in which the self has exploded."[59] Traumatized subjects, "disconnected from their affects," present symptoms analogous to those that accompany brain disorders even though they "are not related to any lesion."[60]

It now appears that the impact of social war is just as forceful as a brain lesion and no less violent than being struck by a bullet or an iron bar. Even if such blows do not always occur as sudden events but tend to be more continuous or harassing, their sense, like that of a brain lesion, remains dissimulated beneath an absence of sense—social conflict without dialectic, as anonymous as a natural catastrophe—an absence that reveals *the very coolness of the political and the social today*.

Damasio affirms:

> Developmental sociopaths or psychopaths are well known to all of us from the daily news. They steal, they rape, they kill, they lie. They are often smart. The threshold at which their emotions kick in, when they do, is so high that they appear unflappable, and are, from their self reports, unfeeling and uncaring. They are the very picture of the cool head we were told to keep in order to do the right thing. In cold blood, and to everybody's obvious disadvantage including their own, sociopaths often repeat their crimes. They are in fact another example of a pathological state in which a decline in rationality is accompanied by a diminution or absence of feeling.[61]

We retain this formulation: "the threshold at which their emotions kick in." Reflecting upon trauma, today, obviously constitutes a reflection on the nature of this threshold. Difficulty letting oneself be touched is the evil of our times, the paradoxical result of being wounded. To be wounded, indeed, is to be touched, struck by a blow—"Touché!", as one says in a duel. To touch thus means precisely to wound (as in *"il a été gravement touché"* [he was gravely touched]).[62] But the "touching" of the wound, today, has generated an inability to feel touching, *an inability to be touched affectively, which is the sign that one has been "touched"—that is, wounded.*

How do we explain this paralysis of touching paradoxically caused by the impact or contact of catastrophe? Davoine and Gaudillière declare: "Whether or not it has a neurological origin, the impossibility of feeling anything blurs the mirror that connects us to ourselves and to others."[63] There is an undeniable link between *brain lesions* and *lesions in otherness*.[64] When there is no organic lesion, "what has been injured is the very dimension of otherness."[65]

What knowledge resides in this coolness? Difficult to explain by recourse to an immanent drive of destruction, impossible to associate with a determinate libidinal investment, the posttraumatic state is a gaping wound that appears on all the battlefields of contemporary society. A normal reaction to an abnormal situation, it inscribes the enigma of its event upon the global stage.

PART THREE

On the Beyond of the Pleasure Principle—
That It Exists

Introduction: Remission at the Risk of Forgetting the Worst

> One can staunch bloody cruelty (*cruor, crudus, crudelitas*), one can put an end to murder by blade, by the guillotine, in the classical or modern theaters of bloody war, but, according to Nietzsche or Freud, a psychic cruelty will always take its place by inventing new resources.
>
> — JACQUES DERRIDA, *Psychoanalysis Searches the State of Its Soul*

At this stage of our analysis, the reader might be expecting the elaboration of a clinical synthesis between psychoanalysis and neurology that explores new therapeutic possibilities. Nonetheless, at the risk of falling short of such expectations, I would like in this concluding part of the book to insist precisely on the dangers of an overly hasty transition toward remission and cure.

Rather than examine in depth the promises of new therapies, I will go farther than before into my analysis of destructive plasticity. Such a decision might seem surprising. But I am convinced that only profound reflection upon destruction, death, and the negativity of the wound will make possible a truly efficacious and pertinent approach to the neuropsychoanalytic clinic. If we do not envisage negative plasticity with sufficient rigor, the distinction between psychoanalytic and neurological causality—that is, the confrontation between sexuality and cerebrality—will remain incomplete.

Is this not what we have already accomplished? In reality, our inquiry is not over. For the moment, all that I have done is to organize a confrontation between the Freudian approach and the neurological approach to events and traumas. I have also distinguished between two logics of psychic destruction. But I have not yet raised the possibility of their *encounter* and thus have not gone beyond the perspective offered by their critical opposition. The question is precisely whether the death drive and the brutal traumatic force of the event can be thought and articulated together at the heart of a new theory that alone could offer neuropsychoanalysis the chance to assume the program that its name opens up.

Formative Plasticity as the Effacement of Destructive Plasticity

As we have seen, Freud does not take destructive plasticity into account— that is, the possibility that new identities are formed in the wake of psychic destruction. The life drives and the death drives, which seem at first to correspond to two incompatible plastic instances—constructive plasticity and annihilating plasticity—never really interfere with the work of a positive and self-regulated originary plasticity. The interweaving of the life drives and death drives is itself *alive*. Freud always elaborates the influx that comes to threaten this self-regulation—that is, the different cases of "disentangling" the two groups of drives—in terms of *elasticity* or *inertia* instead of *negative plasticity*. What exceeds plastic equilibrium is not plastic. There seems to be no middle term between the plasticity of good form and elasticity as the mortiferous effacement of all form. *There is, in Freud, no form to the negation of form*.

Neurology, it is true, insists upon the way in which a brain lesion or traumatic event results in a total transformation of identity. Nonetheless, rather than *theorizing* this type of change as such, it merely *observes* it. Strangely, contemporary neurology has never upheld destructive plasticity as a key concept. It recognizes the phenomenon but not the law. First, no neurologist has explicitly related the current scientific meaning of plasticity to its pathological meaning. There is a certain concept of neuronal plasticity that refers to the positive and constructive sense of the malleable character of synaptic connections. But there is no concept that characterizes the process whereby these connections are destroyed. Second, the formation of identity through destruction is simply invoked as a *morbid eventuality* or a *mere lesional possibility* and not interrogated as a constant *existential possibility* of the subject.

As soon as the motif of an identity formed through destruction is glimpsed, it is abandoned. As if to conjure away the phenomena of destructive metamorphosis that they have just exhibited, all the neurological works that I have cited quickly hasten to discuss cure and remission; or, they hastily return to "good" plasticity—compensatory, reformative, or mending plasticity. Louis Crocq's book, for example, ends with a chapter entitled "Reparation." Judith Lewis Herman's concludes with reflections on "Reconnection."

In concluding chapters such as these, the authors discuss "behavioral therapies," in particular behavioral and cognitive therapies that constitute a mixture of neuropsychiatric and "classical" psychoanalytic practices. Mark Solms, who also concludes his work with considerations of new forms of therapy, speaks of the "neurobiology of the 'talking cure.'"[1] The goal of such therapeutic processes—which combine the talking cure with the development of new adaptive behaviors (new habits and new capacities)—is to *renew cerebral connections.*[2] Such treatments spur new processes of growth that allow the brain to regain its lost plasticity.

It has been proven that cerebral connections undergo change during a "classical" psychoanalytic cure. There is an undeniable link between analytic speech and the facilitation of new synaptic connections.[3] Lisa Ouss, who is part of the Salpêtrière group around Daniel Widlöcher, insists that the encounter between the psychoanalytic and neurobiological approaches can be fruitful. A double reading of a patient's symptoms prescribes "the work of restoring malleability," which, to a certain extent, makes it possible for him to "put himself back together."[4]

These analyses are extremely interesting and deserve to be studied in depth. However, once again, this is not my task. It is more important for me to show that an overly hasty and premature insistence upon therapy, remission, and cure leads to an effacement of the clinical and philosophical problem posed by destructive plasticity.

The Geopolitics of Trauma and Philosophical Questions

The problem is precisely that of the global phenomenon of psychic violence—whether one understands it to be the violence that underlies the suffering of its victims or the violence exercised by criminals and execution-ers. What is emotional coolness today? How should we understand con-temporary sangfroid and indifference?

Such questions necessarily call for philosophical reflection. In *Psychoanalysis Searches the State of Its Soul*, Jacques Derrida calls upon psy-choanalysis to respond to the problem of *cruelty* today. Psychoanalysis pres-ents itself, he says, as the only discourse "that can today claim the thing of psychical suffering as its own affair"[5] and consider the inscription within the psyche of an endless tendency to "make suffer," "suffering just to suffer, of doing or letting one do evil for evil."[6]

Incontestably, the confrontation that we have established between psy-choanalysis and neurology shows that the former is not or no longer—if it ever was—the only discipline that is directly concerned with the fundamen-tal question of the exercise of cruelty. Neurology is no less concerned with this question. It is, indeed, around this question that the two disciplines must encounter one another.

The neurobiological approach to traumas, which takes into account their impact upon cerebrality, leads to a general reelaboration of the question of suffering and wounding—and thus of the question of *evil*. To be ill or to do ill: These are the psychic phenomena whose study does not fall exclusively under psychoanalytic jurisdiction.

It is on this point that we should address Derrida's assertion that the idea of psychic cruelty remains foreign to the positive sciences: "no other dis-course," he writes, "be it theological, metaphysical, genetic, physicalist, cognivist, and so forth—could open itself up to this hypothesis. They would all be designed to reduce it, exclude it, deprive it of sense."[7] Against these assertions, it is important to underscore that neurology, far from being "reductionist," opens, by virtue of its approach to traumatic events, the field of an unprecedented reflection on the permutations of psychic pain in the age of its globalization. In this sense, neurology also challenges philosophy to think this new economy of pain.

When emotional indifference goes beyond every category of neurosis and psychosis to announce itself as the monstrosity of our time, when changes in personality resulting from damage to synaptic connections reveal the plastic and mortiferous power of trauma, when the borders between nature and the social disappear under the impact of cold and neutral brutal-ity, this is the moment when neurology truly challenges psychoanalysis to undergo real metamorphosis. And the challenge is no less forceful even if, as I have shown, neurology paradoxically shares with psychoanalysis a tendency to run from evil toward reparation. It is simply necessary to be less impatient and to radicalize our interrogation of destruction.

Accordingly, after having examined the modalities of this impatience in the name of various categories of positive plasticity (Freud and Damasio), resilience (Cyrulnik), or "monadic illness" (Sacks), I will address my central question: whether it is possible to reconcile *the compulsion to repeat and the formation of new identities through destruction.* I will examine how it is possible to think the conjoined existence, at the heart of psychic destruction, of a *mechanics of repetition* and a *formal creation* that both confirms and exceeds all mechanics.

Is There a Beyond of the Pleasure Principle?

This point leads us finally to examine the problem that underlies the Freudian theory of trauma: Is there or is there not a beyond of the pleasure principle? To this question, Freud ultimately responds in the negative. Nothing, in the end, seems to shake the mastery and authority of the pleasure principle as a fundamental law of psychic regulation. As painful and unpleasurable as it might be, experience of the repetition of trauma or the accident in fact serves to bind (*binden*) an energy that would otherwise purely and simply annihilate the psyche. The compulsion to repeat thus plays the ambiguous role of the mechanical reiteration and the structuring (binding) regulation of suffering. The binding effected by repetition always supports an equilibrium between pure automatism and dynamic order.

What happens if this equilibrium is upset by negative plasticity, which, between two iterations, comes to transform reiteration itself through the *creation of form*? What would happen if the determinism of the compulsion to repeat and the transcendental character of its principle were suddenly *embodied* and came *to form* individual identities and thus *figures of destruction*? Does not the presence of traumatized psychic configurations formed by destruction bear witness to the *palpable and phenomenal* existence of a beyond of the pleasure principle? Of a plasticity irreducible to binding?

Might neurology and psychoanalysis some day work together to rewrite *Beyond the Pleasure Principle* with an entirely new conclusion?

NINE

The Equivocity of Reparation: From Elasticity to Resilience

The problem is quite simple. All we need to do to make it complicated is pose the question clearly. To this effect I shall ask: What is an event? What is this traumatic violence that rips the protective bubble around a person? How does a trauma become integrated into the memory? What is the nature of the scaffolding that must surround the person after the uproar to enable him to resume life despite the wounding and the memory of it?

— BORIS CYRULNIK, *The Whispering of Ghosts*

Freudian Definitions of Plasticity

LIFE AND DEATH, CONSTRUCTION AND DESTRUCTION

The Freudian concept of plasticity, we must recall, essentially designates the imperishable character of psychic life. This character is profoundly ambiguous because it corresponds to two contradictory significations of the term "plasticity": it refers both to the reception and donation of form and to the annihilation of form. On one hand, indeed, the imperishable character of psychic life entails the persistence of the form of memory and its potential reactivation. On the other hand, Freud understands this same character as the paradoxical work of the death drive, a retrograde force that tends to restore a state anterior to life. The imperishable character of psychic life then becomes the paradoxical synonym of a return to the nonliving that annuls any form or any inscription within the neutrality of the inorganic.

One can thus postulate that, for Freud, there are *two plasticities within plasticity*: constructive plasticity and destructive or mortiferous plasticity. In *Beyond the Pleasure Principle*, Freud writes, citing the biologist Ewald Hering: "According to E. Hering's theory, two kinds of processes are constantly at work in living substance, operating in contrary directions [*entgegengesetze Richtung*], one constructive or assimilatory [*die einen aufbauen—assimilatorisch*] and the other destructive or dissimilatory [*die anderen abbauend—dissimilatorisch*]."[1] He affirms that these two directions allow us to recognize "both of our instinctual impulses, the life drives and the death drives."[2] Each drive would be assigned its own physiological process: either construction (*Aufbau*) or decomposition (*Zerfall*).

Construction is formative in the sense that it forges bonds: Eros is a drive of *synthesis* that consists in always multiplying relations between existing units. Death, on the contrary, is fragmentation, the deconstitution of form, *analysis*. In *An Outline of Psychoanalysis*, Freud writes:

> The aim of [Eros] is to establish ever greater unities and to preserve them thus [*erhalten*]—in short, to bind together [*Binding*]; the aim of the [destructive drive] is, on the contrary, to undo connections [*auflösen*] and so to destroy [*zerstören*] things. In this case of the destructive drive we may suppose that its final aim is to lead what is living into an inorganic state [*anorganischen Zustand*]. For this reason we also call it the *death drive*.[3]

The same idea is developed in *Civilization and Its Discontents*:

> Starting from speculations on the beginning of life and from biological parallels, I drew the conclusion that, besides the drive to preserve [*erhalten*] living substance and to join it into ever larger units, there must exist another, contrary drive seeking to dissolve [*auflösen*] those units and to bring them back to their primaeval, inorganic [*anorganisch*] state. That is to say, as well as Eros there was a drive of death. The phenomena of life could be explained from the concurrent or mutually opposing actions of these two drives.[4]

The erotic tendency to bind and to form is present within every living unity, within each cell. Likewise, unicellular beings tend to unite with one another in the course of evolution in order to form complex organisms: "It appears that, as a result of the combination of unicellular organisms into multicellular forms of life, the death drive of the single cell can successfully be neutralized and the destructive impulses be diverted on to the external world through the instrumentality of a special organ."[5] This same plastic movement of union and unity is at work within human societies and marks the origin of civilization. The latter, in fact, "is a process in the service of Eros, whose purpose is to combine single human individuals, and after that families, then races, peoples and nations, into one great unity, the unity of mankind."[6] Against the grain of this constructive tendency, the death drive works to disintegrate these unities.

Freud certainly insists upon the fact that the life drives and the death drives are most often interwoven, that they are "fused [*legieren*], blended [*vermischen*], and alloyed with each other [*miteinander verbinden*]"[7] and almost never appear without one another. It is thus difficult to unmix them, and the constructive and destructive tendencies are liable to exchange roles. But this does not prevent the two groups of drives from very clearly revealing the distinction between their respective dynamics—the one giving form and the other annihilating it.

Why say, then, that something remains unthought within Freud's understanding of these two plastic dynamics? Why affirm that the second— negative plasticity—has not been fully developed and that it has never proven itself apt, in particular, to account for the process of psychic metamorphosis linked to trauma?

THE PLASTIC EQUILIBRIUM OF THE LIBIDO

The term "plasticity" has another determining signification for Freud that concerns the libido's *mobility* (*Bewegenheit*) and *its degree of consistency* (*Beschaffenheit*). The libido is defined as an energy endowed with a specific material tenor (which Freud sometimes calls "substance") that is neither properly a liquid or a solid but represents a middle state between the two. In order to designate this specific quality, Freud will employ the adjective *flüssig* or the substantive *Flüssigkeit*—fluid, fluidity—but, most often, he resorts to the terms *plastisch* or *Plastizität*—plastic, plasticity. Exploring the signification of the terms that are inseparable from the theory of libidinal cathexis and its function leads to the heart of the problem and makes it possible to understand why the negative signification of plasticity never really makes an appearance.

A "plastic" libido is a libido that regulates its cathexes by striking the right balance between attachment—the aptitude for fixating upon an object—and detachment—the aptitude for letting go of it in order to settle upon a new object. The libido is "plastic" when it attaches without solidifying and detaches without dissolving, thereby attaining the right balance between rigidity and inconsistency.

Freud sometimes compares the libido to a unicellular animal—hydra or amoeba—and often describes it using the metaphor of protoplasmic consistency—thicker than water, more fluid than flesh. In "A Difficulty in the Path of Psychoanalysis," he writes:

> For complete health it is essential that the libido should not lose this full mobility [*Beweglichkeit*]. As an illustration of this state of things we may think of an amoeba [*protoplasmatierchen*], whose viscous substance [*zahl flüssige Substanz*] puts out pseudopodia [*Fortsetzungen*], elongations into which the substance of the body extends but which can be retracted at any time so that the form [*die Form*] of the protoplasmic mass is restored [*wieder hergestellt wird*].[8]

It is important for the libido to be able to form attachments even as it remains available for future cathexes: herein lies the condition of psychic vitality.

However, this plastic "happy medium" is predetermined. Freud never clarifies the process whereby the libido might attain this balance. Without explanation, this state is simply identified with *health* and *normality*. Everything that upsets this balance is rejected from the outset as *outside the domain of plasticity*. When the intermediate state between fixation and excessive fluidity has been disrupted, plasticity must give way to other energetic determinations that no longer pertain to its field.

Every reduction of the libido's mobility, in fact, exhausts the concept of plasticity itself, which thereby simply loses its pertinence. Accordingly, when the libido becomes overly fixated, its plasticity gives way to a new "adhesiveness" (*Klebrigkeit*), a "tendency to fixation" (*Fähigkeit zu Fixierung*), a "tenacity" (*Zähigkeit*), or finally an "inertia" (*Trägheit*). Inversely, when the libido becomes excessively fluid, its plasticity gives way to a lability that desubstantializes it.

An example of the first can be found in the case of the Wolf Man:

> Any position of the libido which he [the Wolf Man] has once taken up was obstinately defended by him from fear of what he would lose by giving it up and from distrust of the probability of a complete substitute being afforded by the new position that was in view. This is an important and fundamental psychological peculiarity, which I described in my *Three Essays on the Theory of Sexuality* as a susceptibility to "fixation."[9]

Freud characterizes fixation as a river that has been diverted from its course:

> [T]he libido behaves like a river whose main bed has become blocked [*verhält sich die Libido wie ein Strom, dessen Hauptbett verlegt wird*]. It proceeds to fill up collateral channels which may hitherto have been empty. Thus, in the same way, what appears to be the strong tendency (though, it is true, a negative one) of psychoneurotics to perversion may be collaterally determined, and must, in any case, be collaterally [*kollaterale Wege*] intensified.[10]

This metaphor of the river prefigures the stasis (*Stauen*) of the libido—that is, its rigidification. Diverted into "collateral channels," the libido *stagnates* and *festers* and thus loses all plasticity.

On the subject of the libido that has become overly fluid, Freud remarks:

> One meets with the opposite type of person, too, in whom the libido seems particularly mobile; it enters readily upon the new cathexes suggested by analysis, abandoning its former ones in exchange for them. The difference between the two types is comparable to the one felt by a sculptor according to whether he works in hard stone or soft clay. Unfortunately, in this second type the results of analysis often turn out to be very impermanent: the new cathexes are soon given up once more, and we have an impression, not of having worked in clay, but of having written on water. In the words of the proverb: "Soon got, soon gone."[11]

The comparison with the plastic arts is particularly important to the extent that it clarifies the normativity inherent in the work of the psychoanalyst himself. In the same way that the sculptor can only work upon a material that has a consistency somewhere between polymorphism and rigidity, the psychoanalyst is helpless to do anything with material—that of the psyche—which is either too "hard" or too "soft."

However, just as certain materials are naturally plastic (such as marble or clay), the plasticity of the subject in analysis is itself given in advance and *naturally* determined:

> It may be laid down that the aim of the treatment is to remove the patient's resistances and to pass his repression in review and thus to bring about the most far-reaching unification and strengthening of his ego, to enable him to save the mental energy which is expending upon internal conflicts, to make the best of him that his inherited capacities will allow and so to make him as efficient and as capable of enjoyment as possible.[12]

Nonetheless, the patient's plasticity depends upon his psychic disposition, his age, and what Freud mysteriously calls his "value":

> Since psychoanalysis demands a certain amount of psychical plasticity from its patients, some kind of age-limit must be laid down in their selection; and since it necessitates the devotion of long and intense attention to the individual patient, it would be uneconomical to squander such expenditure upon completely worthless persons who happen to be neurotic.[13]

Or, in the case of the Wolf Man, Freud writes:

> We only know one thing . . . and that is that mobility of the mental cathexes is a quality which shows striking diminution with the advance of age. This has given us one of the indications of the limits within which psychoanalytic treatment is effective. There are some people, however, who retain this mental plasticity far beyond the usual age-limit, and others who lose it prematurely.[14]

Psychoanalysis, therefore, can only work and exercise its sculptural art within the very strict limits of the psyche's plasticity over which it has no control; and so, it finds itself determined by unavoidable natural constraints. Everything that transgresses these limits also transgresses the limits of the concept of plasticity itself. In *Analysis Terminable and Interminable*, Freud writes:

> We are [in certain cases] surprised by an attitude in our patients which can only be put down to a *depletion of the plasticity*, the capacity for change and further development, which we should ordinarily expect. . . . But with the patients I here have in mind, all the mental processes, relationships and distributions of force are unchangeable, fixed and rigid.[15]

This is why, Freud reiterates, "we are . . . prepared to find a certain amount of psychical inertia."[16] The behaviors that he describes are thus deprived of plastic force and the psychoanalyst can do nothing when plasticity has been depleted. *The "depletion" of "good" plasticity has no plastic future.*

The most opaque face of this depletion is that which is described under the name of "negative therapeutic reaction," which indicates the "presence of a power in mental life which we call the drive of aggression or of destruction according to its aims, and which we trace back to the original death drive of living matter."[17] Indeed, this is a "force which is defending itself by every possible means against recovery and which is absolutely resolved to hold on to illness and suffering."[18] But this force—in which the depletion of plasticity and the death drive intersect—is once again defined as an "elastic" tendency, not as a plastic instance of the psyche.

The regulating balance between attachment and detachment is not only operative in the conception of the libido but also determines Freud's conception of plasticity from beginning to end. This is why balance plays such an essential role within the economy of the two psychic tendencies—constructive and destructive—inherent to the drives of life and death. When the "disaggregatory" drive is too strong, it is no longer plastic. When the life drive surrenders to the imperative of binding, it also loses it plasticity. In fact, when Freud speaks of plasticity, he always speaks about it in terms of equilibrium and measure—albeit the equilibrium between construction and destruction, life and death.

It is thus not surprising that the life drives should be presented as *more plastic* than the death drives: "The erotic drives appear to be altogether more plastic, more readily diverted and displaced than the destructive drives."[19] The life drives are obviously more apt to arrive at the best compromise possible between the solidity and liquefaction of desire.

In fact, Freud *never* uses the words "plastic" or "plasticity" to designate the work of the drives of destruction. In *Beyond the Pleasure Principle*, the death drive is said to be "a kind of organic elasticity [*ein Art von organischer Elastizität*], or, to put it another way, the expression of the inertia inherent in organic life [*die Ausserung der Trägheit im organischen Leben*]."[20] However, *elasticity is the exact opposite of plasticity*! While plastic material holds its form and cannot return to its initial state once it has been configured (as happens, for example, with sculpted marble), elastic material does return to its initial form and loses the memory of the deformations that it has undergone. This is why Freud authorizes himself to identify—despite the epistemological incoherence of this correspondence—elasticity and entropy. He affirms that "in considering the conversion of psychical energy no less than of physical, we must make use of the concept of *entropy*, which opposes the undoing of what has already occurred."[21] The loss of form is *entropic*, not *plastic*.

Inertia names the resistance of heavy objects to any movement imposed upon them.[22] Elasticity characterizes the resistance of material to deformation or imprinting.[23] Entropy, according to its etymology (*entropia*), originally signified "turning back."[24] All of these cases—regression, destruction, nullification—are also negations of negative plasticity. For Freud, there is, strictly speaking, no *plastic work of the death drive*.

Nowhere does one find the idea of a plastic autonomy of destruction. It thus becomes clear that, when Freud says that the imperishable character of psychic life is "plastic," he is confusing plasticity (equilibrium, "good" form) and elasticity (nullification of form). Elasticity (or inertia) would thus be the entropy of plasticity—its limit and not its mark.[25]

INCOMPLETE POSITIVE TRANSFORMATION

We must now underscore the profound ambivalence that affects "the problem of preservation in the sphere of the mind,"[26] bearing witness to the "extraordinary plasticity of psychic developments": "In mental life nothing which has once been formed can perish . . . everything is somehow preserved and . . . in suitable circumstances (when, for instance, regression goes back far enough) it can once more be brought to light."[27] To better understand this phenomenon, Freud develops the famous comparison between the psyche and the city of Rome: "We will choose as an example the history of the Eternal City."[28] Freud recognizes that the example is not perfect because only ruins remain of the architectural past of Rome whereas the trace of the psychic past always remains alive. The total preservation of the past, without destruction, alteration, or ruins, is only possible for the psyche: "The fact is that only in the mind is such preservation of all the earlier stages alongside of the final form possible, and that we are not in a position to represent this phenomenon in pictorial terms."[29]

The double, undecidable, and ambiguous sense of such preservation thus becomes crystal clear. It bears witness at once to the vivacity of memory and the amorphous, passive neutrality of nonmemory and inert matter. Indeed, the total preservation of the past comprises not only the preservation of the lived past but also the memory of a past that is older than life. In the first case, preservation is plastic; but, in the second, it is no longer plastic in the strict sense. The *form of life stretches itself out*, retains the memory of its absence of initial form *without giving it form*: it snaps back like an elastic band.

INCOMPLETE NEGATIVE TRANSFORMATION

Plasticity annuls itself in elasticity without creating an intermediary form, without the emergence of an identity formed through destruction. On one hand, positive plasticity is always working over the same material and, from one form to the next, leaves its own remainder: "Portions of the earlier organization always persist alongside of the more recent one, and even in normal development the transformation is never complete and residues of earlier libidinal fixations may still be retained in the final configuration."[30] On the other hand, when this plasticity is depleted, no form remains at all.

Of course, Freud affirms that psychic disturbances transform the ego. The "alterations of the ego" (*Ichveränderungen*) caused by such disturbances correspond to the transformations of "character" that result from events or traumas.[31] There "alterations" even appear to have a certain autonomy:

> All these phenomena, the symptoms as well as the restrictions on the ego and the stable character-changes, have a *compulsive* quality: that is to say that they have great psychical intensity and at the same time exhibit a far-reaching independence of the organization of the other mental processes, which are adjusted to the demands of the real external world and obey the laws of logical thinking. . . . They are, one might say, a State within a State.[32]

This idea that illness forms a state within a state seems precisely to represent the formative work of negative plasticity. But this hypothesis stops short. The "alterations of the ego" produce nothing but scars or, inversely, the disintegration of the ego itself without the intermediate emergence of a new character. Either the "alterations of the ego, comparable to scars, are left behind"[33] or, as the effect of a trauma, "an attempt of this kind . . . ends often enough in a complete devastation or fragmentation of the ego or in its being overwhelmed by the portion which was early split off and which is dominated by the trauma."[34]

Illness does create form. One can consider it as "an attempt at a cure—an further effort to reconcile with the portions of the ego that have been split off by the influence of the trauma and to unite them into a powerful whole vis-à-vis the external world."[35] But this creation is in no way the singular act of forming a new identity.

Neurological Definitions of Plasticity

The contemporary neurological understanding of plasticity suffers from the same shortcomings. The preceding analysis established that neurologists recognize the destructive plastic power of brain lesions resulting from the impact of traumas and catastrophic events upon the psyche. This was the point of view—cerebrality against plasticity—that I upheld in my discussion of Freud.

Nonetheless, it must still be recognized that the explicit elaboration of this negative plasticity is *my intervention*. Never has this concept appeared as such in contemporary neurology. The formation of identity in the aftermath of shock is an established fact that "rehabilitates" the traumatic event and its etiological power. However, much like in Freud, the negative signification of plasticity remains subordinated to the positive signification of giving and receiving form, to the *right balance between suppression and preservation*. Second, there is, in contemporary neurological discourse, a confusion between plasticity and elasticity under the sign of *resilience*. And third, neurology always stresses that *illness is a compensatory creation that functions to repair the world*. No sooner is negative plasticity invoked than it is conjured away.

THE RIGHT BALANCE

It must be recalled that cerebral plasticity has three principal significations: (1) it is the modality of the formation of synaptic connections; (2) it is the modification of these connections effected by individual history and experience; and (3) it is a faculty of compensation and reparation.

In all of its three senses, cerebral plasticity is also situated between *an excess of lability and an excess of fixity*. Antonio Damasio writes:

> Different experiences cause synaptic strengths to vary within and across many neuronal systems, experience shapes the design of circuits. Moreover, in some systems more than in others, synaptic strengths can change throughout the life span, to reflect different organism experiences, and as a result, the design of brain circuits continues to change. The circuits are not only receptive to the results of first experience, but repeatedly pliable and modifiable by continued experience. . . . The idea that all circuits are evanescent makes little sense. Wholesale modifiability would have created individuals incapable of recognizing one another and lacking a sense of their own biography. . . . The brain needs a balance between circuits whose firing allegiances may change like quicksilver, and circuits that are resistant though not necessarily impervious to change. The circuits that help us recognize our face in the mirror today, without surprise, have been changed subtly to accommodate the structural modifications that the time now spent has given those faces.[36]

As this passage makes clear, it is *the right plastic balance* that governs the life of the brain. Constant change would liquidate identity; too infrequent changes would fossilize it. But is there not a destructive plasticity of the happy medium between fluidity and solidification? Wouldn't such plasticity have a role to play? Might it not underlie every scientific description of brain lesions?

The Oxford Companion to the Mind takes a radical position on this point: *There is no destructive plasticity.*

> We do not just mean *any* alteration [when we speak of cerebral plasticity]. For instance, a massive and disorganized malfunction associated with extensive injury would not be referred to as plasticity. To qualify for this name, an alteration has to show pattern or order. Plasticity here means patterned, or ordered, alteration of organization; one that makes some sort of sense biologically or to the investigator.[37]

This definition only retains one signification of plasticity: *structured formation* or *coherent change*. The signification of *formation through destruction* is purely and simply excluded.

RESILIENCE

Can the concept of resilience, which appears widely in contemporary psychopathology, come to the rescue of these "two plasticities"? The problematic of resilience, particularly important to the work of Boris Cyrulnik, seems to take into account the relation between the two significations of plasticity, for it designates the capacity of the subject's identity to reorganize itself (construction) after a shock or a trauma (annihilation of form).

The word "resilience," like "plasticity," comes from the field of material physics. It pertains to the technical vocabulary of metalworking and designates the aptitude of metals to recover their initial form after receiving a blow, being bent, or subjected to continuous stretching or compression. Even as today's new uses of the word retain this physical signification, they also depart from it; for, "resilience" most often designates, not the resistance of inert matter, but that of a complex system. It thus comes to designate the quality that allows systems to regain and maintain their initial functions after being disorganized. It supposes thresholds of tolerance, short of and beyond which the systematicity of the system would be destroyed.

In the field of ecology, for example, resilience characterizes the capacity of an organism, a population, or an ecosystem to recover more or less rapidly from a catastrophe such as a flood, a drought, or a fire. Ecosystems develop many processes of self-regulation and manage to overcome the effects of disorder by progressively reestablishing their initial homeostasis.[38]

Accordingly, the concept of "resilience" ends up designating, in general, the capacity of a functional system to repair itself. Its primary field of application, from that point onward, is psychology. Cyrulnik employs it to characterize, in particular, the situation of vulnerable children who manage to regain emotional equilibrium after experiencing major shocks or abuse. Resilience makes it possible to understand and to evaluate the psychological ability to *bounce back*:

> When the word "resilience" was born in physics, it designated the aptitude of a body to resist shocks. But it attributed too much importance to substance. When it moved into the social sciences, it came to signify "the capacity to succeed, to live, and to develop positively in a socially acceptable manner in spite of a degree of stress or adversity that would normally entail the grave risk of a negative outcome."[39]

Beginning in the 1990s, "the problem of resilience became oriented toward the study of factors of protection"—that is, toward the creation of a secret internal world that allows a victim—in particular, a child—to resist and thus to maintain or even to transform his psychic equilibrium.

Derived from *resilire*, which signifies "jump back" or "pull back," the word "resilience" is very close to *resilement*—that is, nullification or liquidation. Within juridical vocabulary, the verb "to resiliate" means "to annul" or "to retract." However, for Cyrulnik:

> To resiliate an obligation also means to dislodge oneself, no longer to be prisoner to a past. Resilience has nothing to do with some invulnerability or the superior quality of certain people but rather with the capacity to resume a human life in spite of being wounded, without fixating upon this wound.[40]

The process of resilience depends at once on the objective givens (war, genocide, torture, rape, assault, and so on) and the subjective givens (deferred action, reactive decompensation, posttraumatic stress) of trauma. Developing a whole geopolitics of trauma on a planetary scale, relying upon many examples taken from every region of the globe and from different moments in the twentieth century, Cyrulnik considers both sets of givens at the same time and illuminates the precise and always singular relation between the triggering event and its impact on the psyche. Resilience thus appears as "both a synchronic and a diachronic process" that brings "developmental biological forces into articulation with the social context in order to create a self-representation that makes possible the historicization of the subject."[41] Or further: resilience "should not only be sought inside the person, nor within his environment, but between the two, because it ceaselessly weaves processes of intimate becoming into processes of social becoming."[42]

Do these definitions of resilience make it possible to account for the sculptural work of destruction as an effect of a traumatic event, even if this work is preliminary to a reconstruction? In view of the phenomena described here, *is the concept of resilience more satisfying than the concept of plasticity?*

One justification for privileging this term is that resilience describes the point of an event's impact and the phenomenon of resistance that emerges at and from this point. Resilience supposes the irreversibility of psychic time. Afterward, it will never be like it was before. To a certain extent, resilience is the metallurgic equivalent of elasticity. Nonetheless, within the field of physics, the restoration of equilibrium that it implies does not constitute the pure and simple nullification of an initial form, trace, or imprint. Resilience is precisely not a principle of inertia.

On the other hand, resilience is a principle of the *metamorphosis* of identity. The "biological metamorphosis" mentioned earlier corresponds to a change in personality:

> There is no turmoil without metamorphosis. People wounded to the bottom of their souls, the shattered faces of affective emptiness, battered children, and wrecked adults bear stunning witness to the intimate development of a new philosophy of existence. . . . Of such elements resilience is woven.[43]

The triumph of resilience is fragile. The wound is transformed, but it never heals completely. When a subject is severely damaged by existence, he finds himself obliged constantly to uphold the process of resilience to the day of his death. Because trauma is engraved in individual memory, forgetting is never an alternative to healing: "The concept of resilience, which has nothing to do with invulnerability, belongs among the mechanisms of defense."[44] But this does not mean, as people have often averred, that resilience tends to acquit the executioners of their crimes: "The triumph of the wounded never exculpates the aggressor."[45]

The concept of resilience is thus an attempt to show that *something escapes the compulsion to repeat*. A victim is able not to remain fixated upon his trauma.[46] "Repetition is not obligatory," Cyrulnik declares. The example of abused children confirms it: "If it is true that the parents of abused children were themselves often abused children, this does not mean that abused children will necessarily become abusive adults."[47] *A form of life is thus possible that improvises upon and away from compulsion.*

For this very reason, something essential is missing in the theory of resilience; for, ultimately, it remains exclusively attached to the *positive* sense of plasticity. The metamorphosis that it governs is indeed a transformation after a blow, a shock, or a deformation but it always intervenes as a *reconstruction* or a *compensation*. One would never say that the deserted identities discussed in earlier chapters are resilient. Resilience is not compatible with coolness, indifference, or the absence of affect. It never offers *any other option than reorganization*. There is no intermediate position between the resilient personality and the nonresilient personality. The former reconstructs himself, whereas the second perishes without leaving the least room for a third ontological type—the cool resistance of destroyed identity, the persistence of indifference, neutrality as the psychic form of rupture.

OLIVER SACKS AND THE FORMS OF WOUNDED LIVES

Oliver Sacks often lays emphasis upon the nervous system's faculties of reparation and compensation. He would object to my reservations about the concept of resilience on the grounds that *certain compensations are themselves pathological phenomena*. As proof, he would adduce the stunning dimension of "excess" or "superabundance" that accompany certain cerebral pathologies and that make it possible to advance the hypothesis of an organizational pathology, a plasticity that, as it were, *destroys positively*.

In the case of certain aphasias or Tourette's syndrome, for example, we encounter a form of "compensation" or a "sharpening of the compensatory function" through "energetic excess." The second part of *The Man Who Mistook His Wife For a Hat* is precisely entitled "Excesses." By virtue of their excesses, diseases are analyzed as examples of "generative" pathological processes.

These processes produce an intermediate state between health and "deficit." Through disequilibrium, the pathology, in such cases, creates the conditions for a new equilibrium. "What then of . . . an excess of super-abundance of function?" Sacks asks.[48] "Neurology has no word for this—because it has no concept. A function, or a functional system, works—or it does not: these are the only possibilities it allows. Thus a disease which is 'ebullient' or 'productive' in character challenges the basic mechanistic concepts of neurology."[49] The scope of classical neurology thus cannot account for certain supposedly incurable disturbances (such as Tourette's, precisely). In reality, what must be understood is that these disturbances are not simply functional disorders but "hyperactive" economies.[50]

Shereshevsky, the man endowed with a superabundant mnesic function, who became the hero of Luria's narrative *The Mind of a Mnemonist*, suffered from such an "excessive" deregulation. Much like Jorge Luis Borges's character Funes the Memorious, in whom "the prodigious meets the pathological,"[51] this man lives in a world that his prodigious memory will end up fragmenting. Everything is so rich that no synthesis is possible. But this excessive richness, this destructive profusion, still tends toward a form of organization. With the help of the ancient "art of memory"—that is, using a "technique of visual notation that consists in arbitrarily 'placing' clear and precise images . . . in particular locations . . . in his mental structure," Shereshevsky manages to "acquire a sort of inner stability" and becomes a "professional mnemonist."[52]

The damage caused by lesions to one functional system can be compensated for by reorganizing the whole set of systems. Luria declares: "It can be hypothesized that the disorganization of the functional system resulting from a local lesion can be compensated by the reorganization of these functional systems and by inclusion of new preserved links in the damaged system."[53] Claiming the heritage of such pronouncements, Sacks insists, in his turn, upon the necessity of moving "from a neurology of the function to a neurology of action, of life."[54] From the function to the functional system, the entire psychic impact of the wound changes meaning.

The case of Tourette's syndrome, "Witty Ticcy Ray," analyzed in *The Man Who Mistook His Wife for a Hat*, is interesting from this point of view. This syndrome does not correspond to any deficit but rather to an excess of function that, for lack of any lesion, cannot be identified by medical imaging. Rather than consider Ray's energetic superabundance—increased vitality, abnormal responsiveness or reactions, logorrhea—as pure and simple forms of damage, Sacks becomes interested in how these excesses makes it possible for the patient to adapt to the pathology of which they are the nonlesional index. Disease must then be considered as the "ability of the organism to create a new organization and order, one that fits its special, altered disposition."[55] Sacks transposes "compensation," which Luria considered to be a "change of configuration" within the functional systems, to the level of the organism as a whole where it becomes "adaptation" or "reconstruction."[56]

Ray diverts his impulsiveness and his physical tics into game playing and virtuoso improvisations on the drums. In his case, we thus encounter an existence deeply metamorphosed by disease, which produces new ways of reasoning, calculating, and feeling. In order to account for this pathological transformation and for the new order that it produces, Sacks proposes the Leibnizian concept of "monadic diseases." And he brings this concept into relation with the Nietzschean motif of the "perpetual hostility" between health and sickness that unfolds according to the generative process of "overcoming" toward "great health."[57] He explains: "Diseases have a character of their own, but they also partake of our character; we have a character of our own, but we also partake of the world's character: character is monadic or microcosmic, worlds within worlds, worlds which express worlds."[58] Functional systems adapt to damage and organize themselves into new configurations that become worlds. And such worlds correspond to another body and another psyche.

It is thus possible to understand the emphasis that Sacks places upon writing "neurological novels" and upon the necessity of discovering a narrative technique and a rhetoric specific to case histories. Sacks takes up Luria's cherished idea of an opposition between "classical scholars" and "romantic scholars":

> Classical scholars are those who look upon events in terms of their constituent parts. . . . Romantic scholars' traits, attitudes, and strategies are just the opposite. . . . It is of the utmost importance to romantics to preserve the wealth of living reality, and they aspire to a science that retains this richness.[59]

Neurology, then, must become a "romantic science" by defining itself as *neuropsychology*—which would take into account the "biography" of its patients:

> Luria saw his own task (one of his two life tasks) as the refounding of a romantic science (the other being the refounding of neuropsychology, a new

analytical science). The two enterprises were not antithetical, but complementary at every point. Thus he spoke of his need to write two sorts of books: "systematic" books (like *Higher Cortical Functions*), and "biographical" or "romantic" books (like *The Mnemonist* and *Shattered World*).[60]

Accordingly, "the syndrome is always related to the person and the person to the syndrome."[61] Luria apprehends wounded individuals as wholes— "their minds, their lives, their worlds, their *survival*."[62]

There is thus a very close relation between the metamorphosis of an identity that survives with a wound and the story of this metamorphosis—as if the plasticity of writing supported that of systems; as if writing itself repaired the wound that, as it repairs itself, nourishes writing. This structural solidarity between novelistic neuropsychology and its transformed patients demands an entirely new approach to medicine, which Sacks calls "existential medicine." In *Awakenings*, he affirms that "it is the function . . . of existential medicine to call upon the latent will, the agent, the 'I,' to call out its commanding and coordinating powers, so that it may regain its hegemony and rule once again."[63] He even goes so far as to make disease into an essential characteristic of Dasein: "Our health, diseases . . . can only be understood with reference to *us*, as expressions of our nature, our living, our being-here (*da-sein*) in the world."[64] There would thus be *an ontological authenticity of pathological transformation.*

WHAT, THEN, OF COOLNESS?

There is no doubt that Sacks has developed the most profound scientific approach to the link between pathological construction and destruction. The ideas that pathology would be world-forming and that the identity of the patient could be fashioned by the excess or superabundance of the disease itself incontestably situate psychopathology on the way toward construction through destruction. With his concept of "disease world," Sacks accords the traumatic event a power of creation that no one before him—except Luria—had perceived with such clarity and precision. By virtue of his style alone, Sacks has transformed the approach to such transformation, describing hopeless cases in a "romantic" mode.

Nonetheless, Sacks displays a *confidence in disease* that paradoxically but logically upholds his confidence in medicine and therapy themselves. It is significant, in this respect, that Sacks's patients *never cease to feel emotions*.

Even when their emotions have been damaged by lesions or brain dysfunction, the patients always know moments of remission—afforded, for example, by playing music or going to church—when they regain something like intact auto-affection. Their "strange indifference" is always compensated, in one way or another, by differences in behavior specific to each "ego" or to regions of disease-worlds.

Such an approach is respectable and remarkable in that it makes it possible to found a human clinic of neurological disturbances that effaces the boundary between brain and psyche, encouraging therapists always to be *affectionate* with their patients and *affected* by them. Such a conception of medicine is absolutely incontestable. Nonetheless, it does not sufficiently allow for the incursion of the negative. Once again, pathological metamorphosis is always endowed with a coefficient of positivity, of self-recreation, and of world-reconstruction. The destructive signification of plasticity remains in the shadows. Or, at least, it is always oriented toward its redemption or sublation.

How can we think, without contradicting ourselves, a plasticity without remedy?

TEN

Toward a Plasticity of the Compulsion to Repeat

> The difficulty, as far as the death drive is concerned, arises from the fact that it cannot be attributed so precisely with a function corresponding to that of sexuality in relation to the life drives (or love). What we are surer of is that it may alloy with the sexual drive in sado-masochism. But we have also the acute feeling that there are forms of destruction that do not exhibit this fusion of the two drives.
>
> — ANDRÉ GREEN, *The Work of the Negative*

Freud and the Formlessness of the Death Drive

A DRIVE WITHOUT PHENOMENON?

What does it mean to say "without remedy"? I will formulate my thinking as directly as possible. *The limit of psychoanalysis is its failure to admit the existence of a beyond of the pleasure principle.* This beyond, which would also be the beyond of all healing, of all possible therapy, never appears in Freud's text. It does appear in contemporary neurology—but without ever being *thought*.

Despite all his detours, Freud's question—whether there is another principle than the pleasure principle, beyond it, at work within the psyche—receives a negative response. The severe traumatic illnesses envisaged in the text are ultimately considered only as exceptions to the law of the pleasure principle that governs the psyche all by itself—even in catastrophic circumstances that might seem to disregard or transgress this law.

The impossibility of giving form to the beyond of the pleasure principle is part and parcel of Freud's trouble finding a *representative* for the death drive that would do for it what Eros does for the life drives. Once again, for Freud, there is no plasticity of the death drive; it remains *formless, structureless, and figureless*.

Sexuality, as a causal regime, is dualistic. It is constituted both by the sexual and the "other of the sexual," which, according to the law of the dualism of the drives, is linked to the death drive. Without phenomena or representatives, however, it must be recognized that this second etiological instance remains essentially *unenvisageable*.

Freud himself recognizes this fact at the very moment that he begins to elaborate this theory of the life and death drives; since the death drive does not have a representative, the character of its existence is inherently tenuous. "The difficulty remains that psychoanalysis has not enabled us hitherto to point to instincts other than the libidinal ones. That, however, is no reason for falling in with the conclusion that no others in fact exist."[1] André Green notes that in Freud's late theory of the drive,

> Freud no longer speaks of the sexual drive but of the sexual function as a means of knowing Eros, with which it cannot be confused. This representative function does not possess all the properties that Eros does. On the other hand, Freud admits that we do not possess an indication analogous to that which libido represents for the sexual function enabling us to know the death drive in such a direct way.[2]

The figure of Eros does not exhaust all the occurrences of the life drive but it does makes it possible to offer an index or a sensible phenomenon of its multiplicity, whereas there is no such figure that would do the same for the death drive.

SADISM AND MASOCHISM SET ASIDE

In order to discern some "index" for the death drive, Freud turns toward *sadism*. When sadism extricates itself from the sexual drive, it seems well suited to play the role of the representative or the phenomenon of the death drive: "From the very first we have recognized the presence of a sadistic component in the sexual drive. As we know, it can make itself independent and can, in the form of a perversion, dominate an individual's entire sexual activity."[3] If the sadistic drive can, as a partial drive, separate from the libidinal drive, it could then bear witness that the death drive has a degree of autonomy. "If such an assumption as this is permissible, then we have met the demand that we should produce an example of a death drive—though, it is true, a displaced one."[4]

In *The Ego and the Id*, Freud comes to the same conclusion: "The second class of drives was not so easy to point to; in the end we came to recognize sadism as its representative."[5] Indeed:

> Once we have admitted the idea of a fusion of the two classes of drives with each other, the possibility of a—more or less complete—"defusion" of them forces itself upon us. The sadistic component of the sexual drive would be a classical example of a serviceable defusion of the drives; and the sadism which has made itself independent as a perversion would be typical of a defusion, though not of one carried to extremes.[6]

Freud will then extend his search for a representative of the death drive to other phenomena, which, although linked to sadism, are not entirely bound up with it: the drive of destruction and the "drive of mastery" (*Bemächtigungstrieb*), for example, also appear as manifestations of the death drive. In this connection, we might also consider the hypothesis of *primary masochism*.[7]

Nonetheless, as Freud is the first to recognize (which he does in the closing phrase of the preceding citation: "though not of one carried to extremes"), these phenomena or these examples are not entirely satisfying. Of course, "we perceive that for purposes of discharge the drive of destruction is habitually brought into the service of Eros; we suspect that the epileptic fit is a product and indication of a defusion of the drives; and we come to understand that defusion of the drives and the marked emergence of the death drive call for particular consideration among the effects of some severe neuroses."[8] The fact remains, however, that the figures of such defusion—sadism, destruction, mastery, masochism—are all derived from the *love-hate dyad*, from love inverted into hate—that is, once again, from *the intrigue of pleasure*.

> For the opposition between the two classes of drive we may put the polarity of love and hate. There is no difficulty in finding a representative of Eros; but we must be grateful that we can find a representative of the elusive death drive in

the drive of destruction, to which hate points the way. Now, clinical observation shows not only that love is with unexpected regularity accompanied by hate (ambivalence), and not only that in human relationships hate is frequently a forerunner of love, but also that in a number of circumstances hate changes into love and love into hate. If this change is more than a mere succession in time—if, that is, one of them actually turns into the other—then clearly the ground is cut away from under a distinction so fundamental as that between erotic drives and death drives, one which presupposes physiological processes running in opposite directions.[9]

This passage is of capital importance to the extent that it underscores hate's dependency upon love and thereby points to *the inherent incapacity of the death drive to form forms.*

If hate results only from a transformation of love, we must then admit that *the plasticity of love is superior to that of hate.* It is thus not surprising—as we showed in the last chapter—that Freud discovers greater plasticity in the sexual drive than in the death drive. "The erotic drives appear to be altogether more plastic, more readily diverted and displaced than the destructive drives."[10]

If hate itself is nothing but *a form of love*, then it has not been demonstrated that the death drive possesses its own power of formal creation. Once again, it is "constructive" plasticity that triumphs since destruction is conceived as one of the figures of life—transformed and unrecognizable, but still drawn from the apparently inexhaustible reservoir of erotic forms.

This thesis of the incontestable plastic superiority of the life drives—even when it functions to give a form to death through the figures of sadism or hate—is the fundamental reason for the present inability of psychoanalysis to deal in a satisfying fashion with the new economy of suffering, of suffering as making one suffer. The point is not that confidence in the plasticity of the life drives bears witness to some optimism. We know that psychoanalysis has no illusions about the permanent character of cruelty within the psyche. Jacques Derrida was perfectly right to underscore this uncompromising dimension of psychoanalysis. Nor is it a matter of reproaching Freud for softening the problem posed by the existence of the death drive. The question concerns *the form of this existence.*

Very likely, Freud never succeeded in articulating a "representativity" of the death drive that could truly extricate itself in a lasting fashion from life and love—that is, from *sexuality.* Is it really possible to conclude that emotional indifference—the paradoxical but incontestable trait specific to those who suffer and make suffer today—derives entirely from sadism or masochism? There might well be a link between them, but it remains less certain that the one proceeds from the other. To be permanently traumatized: not to want anything anymore, not to feel anything, to be prostrated, to have lost all affect; or, inversely, to kill in cold blood, "to blow oneself up," to organize terror, to endow terror with the face of a fortuitous event, to void it of sense—is it really possible, once again, to explain these phenomena through sadism and masochism? Is it not evident that these phenomena have a source other than the transformation of love into hate or of hate into an indifference to hate—a source in a beyond of the pleasure principle endowed with its own plasticity, which it is now time to conceptualize.

The Ambiguity of Repetition in Beyond the Pleasure Principle

COMPULSION FIRST APPEARS AS THE OTHER OF PLEASURE

Let us return once again to *Beyond the Pleasure Principle* and recall that the central task of Freud's 1920 work is to question the authority and unicity of the pleasure principle: is it the sole and original principle governing the economy of the psyche? "In the theory of psychoanalysis," Freud recalls in the opening lines of the work, "we have no hesitation in assuming that the course taken by mental events is automatically regulated by the pleasure principle."[11] Can this authority be contested? Are there unpleasurable experiences that tend to undermine it? Recalling a number of such experiences—in particular, the transformation of pleasure into unpleasure by way of repression—Freud strangely argues that none of them is really capable of calling the domination of the pleasure principle into question: "This does not seem to necessitate any significant limitation of the pleasure principle" other than the limitation, without real significance, which seems to occur in cases of "pleasure that cannot be felt as such."[12]

Nonetheless, as Freud recognizes at the end of the first chapter, "the investigation of the mental reaction to external danger is precisely in a position to produce new material and raise fresh questions bearing upon our present problem."[13] This is the point at which he begins to discuss accidents and catastrophes "involving risk to life" along with other traumas previously mentioned; and, from there, to uphold the existence of the compulsion to repeat that seems to fixate the patient to his trauma and to vie with the pleasure principle; for, what recurs in the patient's dreams is precisely the painful experience of trauma which the overly restrictive categories of pleasure and unpleasure cannot assimilate.

Freud then writes: "We are therefore left in doubt as to whether the impulse to work over in the mind some overpowering experience so as to make oneself master of it can find expression as a primary event, and independently of the pleasure principle."[14] Dreams that lead patients back to the situation of their accidents or even the *fort/da* game whereby the child reiterates the absence of his mother lead Freud to conclude that "the compulsion to repeat also recalls from the past experiences which include no possibility of pleasure, and which can never, even long ago, have brought satisfaction even to instinctual impulses which have since been repressed."[15]

Is the compulsion to repeat then more originary than the pleasure principle and also independent of it? Does it call into question its dominance over psychic life? Chapter 4, which defines trauma as something that pierces the "protective shield" (*Reizshutz*), seems to formulate a positive response to these questions. The psychic apparatus does not have time to prepare for an effraction and finds itself overwhelmed by the violence of trauma—violence that the homeostatic function of the pleasure principle has neither the time nor the power to liquidate.

> At the same time, the pleasure principle is for the moment out of action [*ausser Kraft gesetzt*]. There is no longer any possibility of preventing the mental apparatus from being flooded with large amounts of stimulus, and another problem arises instead—the problem of mastering the amounts of stimulus which have broken in and binding them, in the psychical sense, so that they can be disposed of.[16]

The psychic apparatus, in cases of trauma, no longer seeks pleasure but, instead, busies itself with "binding" the quantities of excitation submerging it. "Binding" (*Bindung*) thus makes its entrance at this point. To bind, for Freud—who says that he has "taken into consideration Breuer's hypothesis"—consists in transforming freely flowing or "unbound" energy into "quiescent cathexis." "We may perhaps suspect that the binding of the energy that streams into the mental apparatus consists in its change from a freely flowing into a quiescent state."[17] We are then given to understand that repetition and binding work together to domesticate traumatic energy. Accident dreams, for example, work "with a view to the psychical binding of traumatic impressions" and thus "obey the compulsion to repeat."[18] But this domestication or taming does not produce pleasure. On the contrary, they do nothing but whet the agonizing process of machinal iteration. It thus becomes clear that compulsive dreams constitute "an exception to the proposition that dreams are fulfillments of wishes."[19]

BUT THERE REMAINS A BINDING AGENT . . .

Following these developments, Freud presents the dualism of the life and death drives in Chapters 5 and 6. This presentation logically follows that of the compulsion to repeat, and Freud attempts to show the indissoluble link between repetition and the death drive: "We still have to solve the problem of the relation of the instinctual processes of repetition and the dominance of the pleasure principle."[20]

Nonetheless, the more complex the analysis becomes, the less it succeeds in making the compulsion to repeat coincide with the death drive. The ambivalence that structures Freud's entire argument derives from the inherent ambivalence of the processes of repetition and binding themselves. To the very end of the book—that is, until Chapter 7—where Freud clearly addresses this problem, we do not know whether repetition and binding escape the pleasure principle or whether they ultimately lead back to it, as two exceptions that prove the rule.

The latter hypothesis increasingly prevails. Compulsive automatism is certainly designed, through repetition and binding, to regulate traumatic influxes of energy that constitute a real danger for the psyche and that the pleasure principle alone cannot manage. Repetition is the default form of homeostasis. Whereas the pleasure principle effortlessly obtains stability, the work of repetition is never complete. But the result remains the same: homeostasis, quiescence, or tranquility.

Chapter 7, which concludes the work, confirms what we had begun to suspect: as repetition binds free flowing energy, it opens up the field of the pleasure principle so that it may continue to exercise its authority. It prepares the ground by tranquilizing energy. However, this process of preparation, which might appear to be the work of a more originary principle than the pleasure principle, remains *an aspect of its work.*

Binding is one of the functions of the pleasure principle. The work thus closes upon this stunning analysis:

> We have found that one of the earliest and most important functions of the mental apparatus is to bind the instinctual impulses which impinge on it, to replace the primary process prevailing in them by the secondary process and convert their freely mobile cathectic energy into a mainly quiescent (tonic) cathexis. While this transformation is taking place no attention can be paid to the development of unpleasure; but this does not imply the suspension of the pleasure principle. On the contrary, the transformation occurs on behalf of the pleasure principle; the binding is a preparatory act which introduces and assures the dominance of the pleasure principle. . . . The binding [once again, impossible without repetition] of an instinctual discharge would be a preliminary function designed to prepare the excitation for its final elimination in the pleasure of discharge.[21]

We might ask, however: if binding prepares the way for the pleasure principle, does this constitute proof that the one is more originary than the other and must be situated *beyond* it? Not at all. This is how Freud resolves the question of the propaedeutic status of binding:

> Let us make a sharper distinction than we have hitherto made between function and tendency. The pleasure principle, then, is a tendency operating in the service of a function whose business it is to free the mental apparatus entirely from excitation or to keep it as low as possible.[22]

The tendency would thus be "older" than the function, even if the function precedes it within the order of time.

Once again, what is at stake in the process of binding is pleasure. Pleasure is always the result of lowering the quantity of excitation; and this is precisely what binding accomplishes when it transforms mobile energy into immobile or bound energy. This low energetic level resembles the inorganic state toward which the death drive strives. But, of course, it also resembles pleasure.[23] There is thus no beyond of the pleasure principle—only a pleasure that must first be bound before it can gain control over the psychic apparatus. The anteriority of this "before" is simply chronological and not hierarchical. Binding through repetition is merely a phase. There is a beyond or a short of the pleasure principle—but *it only abides for a short time* as a kind of preface to the pleasure principle. Trauma forces binding to write this preface to pleasure, but it can never determine what comes afterward. Very quickly, the pleasure principle regains its supremacy.

The profound ambiguity of repetition thus inheres in his binding power. Certainly, in an essential respect, this power is mortiferous: it immobilizes, freezes, or leads to inertia and to the inorganic state. Compulsion—as has been said and resaid—has the spectral character of a death machine. At the same time—something said less often—this "mechanicity" is *a binding agent*: it disciplines, flattens, and tames as it immobilizes. Freud states as much in explaining that accidents are not the fulfillment of wishes: "We may assume, rather, dreams are here helping to carry out another task, which must be accomplished before the dominance of the pleasure principle can ever begin."[24] Such dreams thus "afford us a view of a function of the mental apparatus which, though it does not contradict the pleasure principle, is nevertheless independent of it and seems to be more primitive than the purpose of gaining pleasure and avoiding unpleasure"[25]; and this function is only a propaedeutic moment that ultimately works in the service of pleasure and very quickly loses its status as a supplement.

The ambiguity of repetition thus resides in the unifying value of binding that is inherently permeated by the synthetic value of Eros. Above all, binding *immobilizes the plasticity of dead energy*. For Helmholtz, "free energy" was defined as "that energy which 'is capable of being transformed into other sorts of work,' " while "bound energy" was "the kind 'which can only manifest itself in the form of heat.' "[26] *Free* energy is transformable; *bound* energy can no longer be transformed. Binding exhausts the plasticity of free energy, restricts this energy's power of formation and transformation, and thereby obliterates the hypothesis of a plasticity of death. Once immobilized, energy is returned to inertia. Binding plasticity, once again, gains the upper hand. Compulsive rigidity melts into the suppleness of the binding agent.

In order for Freud positively to conclude that there is a beyond of the pleasure principle, he would have had to admit, or demonstrate, that a wholly other psyche than the one regulated by the pleasure principle is liable to emerge in the aftermath of trauma or catastrophe. He would have had to recognize that the psyche has a specific form produced by the presence of death, pain, or the repetition of unpleasurable experiences. He would have had to do justice to the accident's force of existential improvisation and to psyches deserted by pleasure who remain psyches even though indifference and detachment have vanquished binding.

When Freud speaks of a representative for the death drive, he is seeking the form of this drive. He never finds this form, however, because he deprives destruction of its own plasticity. Sadism and masochism—which, once again, do not purely and simply inhabit a space beyond the pleasure principle—are quite far from such living figures of death.

THE DETERMINIST CHARACTER AND THE PLASTIC CHARACTER OF THE COMPULSION TO REPEAT ARE NOT OPPOSED TO ONE ANOTHER

We must then acknowledge that there is *a plasticity of the compulsion to repeat.* This assertion requires not so much that we call into question but rather that we reconceive the determinist—regular and necessary—character of the compulsion. Likewise, I do not align my rereading of repetition under the sign of difference. It is not a matter of claiming that repetition can produce something other than the identical and thus escape the character of its own law. In attempting to elaborate something like the form of the death drive, I do not intend to show that repetition itself possesses a plastic power that would divert it from its own trajectory, a suppleness that would lead it to open gaps and produce singularities. What I do wish to underscore is that reiteration entails a power of transformation, and that this power is defined less by the production of differences than by *the possibility of not binding repetition.*

The plasticity that I am attempting to think introduces a scission between repetition and binding; it interrupts the synthetic power of binding, cohesion, and discipline. Plasticity, in this sense, does not deprive repetition of its work, but rather forms a disconnection between repetition and binding by distancing repetition from that which it repeats. Without undue play on words, but also without being able to avoid it at this conjuncture, I would say that *Indifference and Repetition* is the book that needs to be written today, showing that destructive plasticity is precisely what renders the psyche indifferent to its own compulsion.

Much like the narcissistic hypercathexis in cases of organic wounds, binding, for Freud, remains a healing power. It constitutes the reparative dimension of repetition. Traumatized identity does not really have the time to pierce a fabric that has already been patched up with pleasure. If there is a beyond of the pleasure principle, on the other hand, it would have to be sought there where the psyche breaks off, on the side of a leap out of the meshwork and continuity that holds it together, or, indeed, of a hole, where the psyche would be severed from all bonds without thereby being delivered over to a destructive pleasure (albeit that of a "primary process"). This breaking off would tear the psyche away from both love and hate. There are psyches that are beyond both love and hate, without being either sadist or masochist.

The plasticity of compulsion does not change its determinism. All traumas tend to repeat themselves and this law remains true. We must simply admit that there is a power of formal creation at the very heart of the necessity of the law. Creation of the faces of the law itself: coolness, disaffection, and indifference—these forms owe nothing to the metabolism of Eros.

Figuring the Beyond of the Pleasure Principle

COMPULSIVE MODIFICATION

How to think this rupture with erotism, albeit the erotism of hate, this dis-intrication that forms the figures of negativity? This is the precise point at which cerebrality comes into play. The appearance of figures of trauma within contemporary neurology, the upsurge of phenomena of haunting and death in cases of PTSD, and the entry of coolness and disaffection onto the scene of worldwide psychopathology, authorize us to affirm that a beyond of the pleasure principle is manifesting itself and taking form. These phenomena go beyond psychoanalysis even as they force it both to rearticulate and consolidate its thinking of the death drive.

The beyond of the pleasure principle would thus be the work of the death drive as the formation of death in life, the production of individual figures that exist only within the detachment of existence. These forms of death in life, freeze-frames of the drive, would be the "satisfying" representatives of the death drive that Freud sought for so long so far from neurology.

This *phenomenology of destruction*, inseparable from the recognition of the rupture between sexuality and cerebrality, delineates the political field of another understanding of suffering and making suffer than the one that is based on sadism and masochism.

The idea of a plasticity of destruction—which makes it possible to *embody* the death drive itself—implies, in accord with the now well-known definition of plasticity, both the reception and the production of form. The trauma-tized psyche is initially a psyche that receives a shock and its formation proceeds from this very reception. The cruelty of executioners—*active* destructive plasticity—thus emerges as the mimetic reappropriation of traumatic passivity. "Making suffer" today manifestly assumes the guise of the neutrality and senselessness of a blow without author and without his-tory, of mechanical violence, and of the absence of interiority—thereby adding another valence to the inexhaustible concept of the banality of evil. Evil is the becoming insensate of evil.

This is the way in which destructive plasticity reveals the possibility, inscribed within each human being, of *becoming someone else at any moment.* Let us remember the case of Phineas Gage and the conclusion that anyone who has suffered the same wound would have suffered the same change of identity. But let us also think of those who make suffer and who, to that end, make themselves other, in accord with the decision of a metamorphosis that is itself also constituted on the level of the brain as indifference to suffering.

The constant risk of destructive metamorphosis should not be identified with the risk of death. It is not only my death that is possible at each instant, as Heidegger says, but also the destructive transformation of the ego. Cerebral auto-affection can, at any instant, be interrupted. The fact that the transformation of self by a wound can be at once destructive and transfigur-ing is the great lesson of contemporary neurology. It is also its political message.

"THEY HAVE ALL BEEN DEAD"—OR HOW SPINOZA WAS RIGHT

It remains necessary, in order for this message to be heard, that neurology itself recognize the work of destructive plasticity and stop covering it over with the curative or compensatory virtues of "good" plasticity. Destructive plasticity is *beyond the principle of plasticity* (if we understand this principle to name the constructive value of plasticity, the only value that the neurobio-logical concept of plasticity has hitherto taken into account). We must then cease considering destructive plasticity as a merely accidental interruption of constructive plasticity and come to see in it a true *cerebral possibility*, which can always part company at any moment with constructive plasticity, *discon-necting* from it.

What happens when we escape trauma? What proportion of the wounded will be revisited by the horror we thought they had overcome? These questions call for studies covering the entire life cycle.

  But they have all been dead, even those who returned home smiling. They have all been in the arms of the unnameable aggressor: death itself "in person." We can go on living afterwards, even laugh about it when we come back from hell, but we hardly dare to admit that we feel initiated by the terrible experience. When we have lived among the dead, when we have lived through death, how can we say that we are ghosts? How can we make others understand that suffering is not depression, and that often the return to life is what hurts?[27]

The figures of the death drive, the plastic creations of destruction, are thus less figures of those who will die than those who are already dead; or rather—resorting to a strange and terrible grammatical contortion—those who *have already been dead*, who "have lived" death, as Cyrulnik says. Those who have come back—as in the Orpheus myth that we evoked—are those people who, in the very form of their psyches, present us with an image of the place they have come from, an image that does nothing but prolong their death in the past (or the "deathlike break" that Cathy Caruth has evoked).[28] The figural improvisation that emerges within the experience of death is the phenomenon that neurology must adopt as its defining question without occulting it through confident insistence upon reparative plasticity. This is not a matter of negating such reparative power, but rather, on the contrary, to offer it a true chance; for, in order to heal the other, one must first identify his sickness.

Spinoza avait raison ("Spinoza was right") is the title of the French translation of Antonio Damasio's book *Looking for Spinoza*. Spinoza was right about the relation between the soul and the body, the author affirms. Certainly. But the philosopher was also right about something else which seems to have eluded the neurologist. Spinoza was right to affirm, in the scolium to proposition XXXIX of his *Ethics*, that a man can die before being dead, that he can undergo destructive changes that render him unrecognizable. This annihilating metamorphic power is as difficult to think as death:

Here it should be noted that I understand the body to die when its parts are so disposed that they acquire a different proportion of motion and rest to one another. For I dare not deny that—even though the circulation of blood is maintained, as well as other [signs] on account of which the body is thought to be alive—the human body can nevertheless be changed into another nature entirely different from its own. *For no reason compels me to maintain that the body does not die unless it is changed into a corpse. . . . And, indeed, experience seems to urge a different conclusion. Sometimes a man undergoes such changes that I should hardly have said he was the same man.*[29]

ELEVEN

The Subject of the Accident

> What we have learned about the structure and functioning of the cerebrospinal apparatus means that we must change the way in which we represent the mechanism of subjectivity in general—hence, the displacement that the notion of the cerebral unconscious seeks to register.
>
> — MARCEL GAUCHET, *L'inconscient cérébral*

Who, today, is this modifiable and metamorphosable subject, the site of conflict between the two plasticities—constructive and destructive—that entwine and menace its life? Before concluding, I will attempt to outline the theoretical and philosophical framework that makes it possible to glimpse this subject.

I said, at the beginning of this study, that in large measure "continental" philosophers have nothing but contempt for the "cerebral" subject. None see in it the future of the subject, the future of the very concept of the subject, while, as Gauchet has argued, it has become increasingly obvious that neurobiological discoveries are intimately concerned with "the idea of human functioning in general, which includes subjectivity."[1]

Another Chapter of The History of Sexuality?

It would certainly be possible to conclude that this reconception of the subject, made possible by neuroscientific advances, is nothing but a new episode of what Michel Foucault has analyzed as the relations between the subject and truth. Contemporary neurobiological research, then, would contribute nothing new and would only be a further episode within the history of "subjection." The category of cerebrality could be reduced to and counted among the "discursive productions" or the "effects of power" that lead one to formulate the truth of the subject at a specific moment in its history.[2]

If it is true that "power passes within the materiality of bodies," if it is true that we now live within the sphere of "biopower and somato-power," then the preponderant place granted to the discourse on the brain within the contemporary global scientific and cultural horizon only amounts to a new modality of the disciplinary techniques that produce the subject as they normalize it. The brain would emerge as the contemporary form of subjectivity only "because relations of power had established it as a possible object."[3]

In the present book, I decided not to rehash the analysis of the coercive ideological power of contemporary neurobiological discourse that I undertook in *What Should We Do with Our Brain?* I have not addressed the fact that the brain today has become the site of the truth of the subject and, as such, its organization has emerged as the dominant sociobiological model for thinking and regulating every systematic configuration— from business to social relations in general and, of course, to the form of bodies.

One might object that my method in this book of redefining the brain in terms of affective economy, showing how cerebrality is constituted as an etiological principle, constitutes a displacement of the concept of sexuality that does nothing to change the meaning of this concept, for such a displacement would remain inscribed within the order of the "erotics of truth"—that is, within a "mixed regime of pleasure and power."[4]

The Foucauldian critique of the way in which psychoanalysis constitutes sex as a site of truth is founded precisely upon a refusal to identify pleasure and sexuality: "The rallying point for the counterattack against the deployment of sexuality ought not to be sex-desire, but bodies and pleasures."[5]

However, the recognition that there is a freedom of pleasures independent of sexual development as Freud described it effectively stripped sexuality of its *autonomous* explicative power. Another body appeared that would not and could not be folded into this interpretative matrix. But it remains uncertain whether upholding cerebrality as an etiological principle against sexuality functions in the same way to liberate the body. From a Foucauldian point of view, *cerebrality would be yet another avatar of sexuality*, a normative power aiming to organize new emotions by giving them a natural basis. In this sense, the present book would simply be the propaedeutic to a new chapter of *The History of Sexuality*.

Nonetheless, I think that situating subject *beyond the pleasure principle* also obviously places the subject *beyond the will to know*. This would be the case for at least two reasons.

First, a genealogical critique of cerebrality is undoubtedly possible, but it avoids the question of what genealogy itself, in its structure, owes to the *neuronal model*. Readers of Foucault might not have sufficiently taken into account the fact that neuronal functioning, organized in centerless networks that depend upon ever-changing punctual interactions, is precisely the biological counterpart to the schema of micropower. The very form of the genealogical critique of biological power would thus be derived from one of the powers that it examines: that of *cerebral organization*.

In "Nietzsche, Genealogy, History," Foucault describes Nietzschean genealogy as a force field. Vectors of power "are displayed superimposed or face-to-face . . . the space that divides them . . . the void between them."[6] This model of the field of forces and counterforces is precisely what is at work in cerebral functioning. It is time to recognize that, for at least half a century, the form of political critique largely follows the very neuronal model that it sometimes attempts to deconstitute.

Second, to refuse to inscribe cerebrality within the framework of *The History of Sexuality* makes it possible to discover that the subject beyond the pleasure principle coincides with the disappearing subject that Foucault discerned in the figure of the writer, the author, or the nondisciplined body. To bring to light a psyche that is so vulnerable to being wounded that it can undergo transformation without retaining any trace of itself, to think a subject who becomes the very form of its death, who, through the interruption of his affects, figures his own disappearance, is to discover, within the revelations of today's neurology, the *material* image of the disappearance of the author. Accordingly, the neuronal is not the other of the genealogical as Foucault thinks it, but rather its mirror image.

In his famous essay, "What Is an Author?" Foucault analyzes the figure of the contemporary author as an evanescent figure. He recalls Beckett's words—" 'What does it matter who is speaking,' someone said, 'What does it matter who is speaking' "—and then comments: "In this indifference appears one of the fundamental ethical principles of contemporary writing."[7] Is the indifference of the subject of writing akin to the emotional indifference of the traumatized subject who has gone beyond the pleasure principle? Writing, Foucault adds, "linked to sacrifice, even to the sacrifice of life; it is now a voluntary effacement that does not need to be represented in books, since it is brought about in the writer's very existence."[8] "There is no one when I write": Isn't this to say, with the new wounded, "there is no one when I live?"

Accordingly, rather than critique cerebrality from a hermeneutic or genealogical viewpoint, wouldn't it be more interesting and more urgent to place the motif of cerebral desertion into relation with that of the disinheritance or deconstruction of subjectivity? Isn't it time that philosophy discover the cerebral psyche as *its subject*? Isn't the subject of cerebrality the sacrificial witness of the philosophical subject?

What Does "Beyond" Mean?

"Deconstruction," properly speaking, has remained strangely closed off to these questions, even though they are directly relevant to it. It is difficult not to notice Jacques Derrida's hostility to the theoretical or philosophical pretentions of contemporary neuroscience. Without further examination, he aligns "neuronal sciences" with "analytic communities" and "positivist or spiritualist models."[9] Never does he envisage their interaction with deconstruction itself or even any other type of philosophy other than so-called analytic philosophy. Nor does he ever identify them as conditions of the future of psychoanalysis in the very lecture, entitled "Psychoanalysis Searches the State of Its Soul," where he attempted precisely to delineate this future.

The central question of this lecture, let us recall, is the question of cruelty and the capacity of psychoanalysis to confront the contemporary faces of evil. Derrida's condemnation is unswerving:

As I see it, psychoanalysis has not yet undertaken and thus still less succeeded in thinking, penetrating, and changing the axioms of the ethical, the juridical, and the political, notably in those seismic places where the theological phantasm of sovereignty quakes and where the most traumatic, let us say in a still confused manner the most cruel events of our time are being produced.[10]

Likewise, psychoanalysis would have taken no clear position on the urgent questions that trouble the new scene that

since the Second World War has been structured by unprecedented juridical performatives . . . such as the new Declaration of Human Rights—the rights not just of man, as we say in French, but of woman as well—the condemnation of genocide, the concept of crime against humanity . . . the creation under way of new international penal authorities, not to mention the growing struggle against the vestiges of forms of punishment called "cruel" . . . namely, besides war, the death penalty, which is massively enforced in China, in the United States, and in a number of Arab Muslim countries.[11]

Such analyses are certainly incontestable, but it remains surprising that the notion of *traumatic event* is never recalled to its primary meaning as psychic wound. Derrida thematizes geopolitical changes in cruelty without ever envisaging psychic change or, more precisely, the change of psyche—that is, *the unconscious*—that corresponds to it. Psychoanalysis is called to respond all alone—using its own forces and its own concepts—to the contemporary challenge of cruelty. "If psychoanalysis does not take this mutation into account . . . it will itself be . . . left on the side of the road."[12]

But how could it take into account contemporary "traumatic events" if these very events, as well as the new psychic structures of trauma, have already exceeded it? How can we explain Derrida's silence on the neurological definitions of trauma, definitions without which psychoanalysis could not advance nor enter into the scene of psychic pain?

"We await today a new 'Thoughts for the Times on War and Death' (I am citing some titles of Freud: "Zeitgemässes über Krieg und Tod" [1915] and a new "Why War?" ["Warum Krieg?" 1932]), or at least new readings of texts of this sort," Derrida pursues. We must rethink "the concept of front, the figure of a front line . . . or a capital front indissociable from that of war."[13] But how is it possible to speak of the front or of the head without evoking the brain and everything that war psychiatry could teach us about the psychic effects of conflicts? How can we not see that the new front line today passes through the very concept of the unconscious? And how could we ignore the neurological sense of the *frontal*?[14]

According to Derrida, it is by returning to the profound sense of the death drive—beginning with the drive of power or mastery—that psychoanalysis can and should confront at renewed expense the configuration formed by the alliance of sovereignty and cruelty. It is a matter of interrogating the contemporary form of the death drive and discovering the "new forms of cruelty that a psychoanalyst of the year 2000 would have to interpret at renewed expense, outside or within the institution. With regard to the political, the geopolitical, the juridical, the ethical, are there consequences, or at least lessons to be drawn from the hypothesis of an irreducible death drive."[15]

Such assertions compel us, once again, to question the possibility of making the unprecedented emerge from the old. Indeed, in Derrida's discourse, the death drive remains what it has always been—that is, the instinctual foundation of the forms of destruction that are "sadism . . . a ferocity that the narcissistic libido would have detached from the ego so as to train it on an object—unless it is that of a primary masochism, a hypothesis that Freud also retains."[16]

The possibility that there could exist figures of the death drive other than sadism or masochism or that the plasticity of this drive could work *beyond love and hate*—such possibilities are never considered. Accordingly, if the phenomena of the death drive are not defined in a new way, how could one expect to grasp its renewed importance within geopolitics today? Such a renewal could only occur by chance—and this is even what Derrida seems to postulate by insisting upon the unanticipatable or unforeseeable event that would make it possible for psychoanalysis today to *hold the front line*. Psychoanalysis will make it through *if it is lucky*.

In order to understand Derrida's position, we must return to the structure of the "beyond" as Freud sets it up in *Beyond the Pleasure Principle*. The task of psychoanalysis, from now on, should consist in engaging with the possibility of a "beyond" of the "beyond of the pleasure principle," a "beyond" that should likewise become *a beyond of the death drive*. The ethics of psychoanalysis would revolve around the regulative ideal of a "beyond of cruelty." "Would there also be, a few steps further beyond the principles, a beyond of the beyond, a beyond of the death drive and thus of the cruelty drive?"[17]

Such a beyond, once again, could only take the form of an unforeseen and unanticipatable event that entirely escapes the logic of the drive itself.

This "beyond of the beyond" of the death drive would be absolutely irreducible to the determinist character of the drive—which is precisely why it can open beyond it. The event of this beyond, Derrida writes, is "like everything that happens, like every event worthy of the name, like everything that is coming, in the form of the impossible, beyond all convention and all scenic control, all pleasure or reality principle, beyond all drive for power and perhaps all death drive."[18]

These declarations call for a remark about the notion of the "beyond" itself. I have attempted to show that Freud ultimately denied the existence of a beyond of the pleasure principle. At the end of *Beyond the Pleasure Principle*, Freud dismisses the hypothesis that certain traumatic events could operate according to a logic that would exceed that of the pleasure principle—such as the logic of the compulsion to repeat. Strangely, it turns out that repetition itself works in the service of the binding that readies trauma to be assimilated by the psyche, which, in the last instance, remains governed in every case by the pleasure principle. That said, how could we envisage a "beyond of the beyond" if there is no beyond of the pleasure principle?

If Freud fails to discern such a beyond, it is precisely because he never manages to think the way in which the death drive might form its own figures. The representatives that he privileges remain sadism and masochism, two phenomena that—no matter their violence or their degree of cruelty— always remain tributary to the pleasure principle, to the enjoyment linked to "making suffer." However, once again, how could we envisage going beyond a beyond that does not exist?

Only a certain neurological approach to psychic destruction can body forth this beyond that, for Freud, remains ever in abeyance. More precisely, it is possible to extend this neurological approach to propose a concept of destructive plasticity—between neurology and psychoanalysis—that endows the death drive with its own (autonomous) form. The deserted identities of cerebrality, living figures of death, emerge precisely as representatives of the tendency to annihilation and destruction that psychoanalysis has always failed to bring to light.

But recognizing this fact supposes the acknowledgment that the future of psychoanalysis does not depend upon psychoanalysis *alone*. Without a serious and constructive dialogue with neurology that would primarily address the sense of a beyond of the pleasure principle, it is difficult to see how psychoanalysis, relying exclusively on its own forces, could achieve some new state of consciousness that could lend old concepts—such as sadism and masochism—the illusory appearance of a second childhood.

Miraculous events can always happen! Even so, why should the event conceived as "absolute arrival" be envisaged outside of any concrete determination of what constitutes an unanticipatable event—such as brain lesions, damage, or trauma? Isn't it precisely by accepting such events, which no longer fall under the jurisdiction of sexuality, that psychoanalysis can finally put itself in a position to flesh out the death drive, the beyond of the pleasure principle, and a new regime of events?

It is hardly surprising that deconstruction has never considered the fact that the contemporary neurological revolution—far from being a mere "scientific phenomenon" that threatens philosophy without being of any philosophical interest—is perhaps capable of offering a concept of the event that no longer owes anything to the traditional conception of the accident and of damage. Isn't cerebrality the fulfillment of the deconstruction of subjectivity?

Conclusion

In *Beyond the Pleasure Principle*, binding is the most important function of the psychical apparatus, which binds the destructive external quantities of excitation in order to master them, even before the intervention of the pleasure principle. Binding is thus the mechanism that serves to protect the organism against the unpleasurable unbinding of the ego caused by excessive stimulation, or trauma. . . . By binding excitations, the organism defers its own death drive. Binding carries also an explicitly political meaning: by binding or bonding the individual with the other or outside in an emotional bond of identification that constitutes the homogeneous group or mass, individuals neutralize their lethal tendency to disband into a disorderly panic or all against all.

— RUTH LEYS, *Trauma: A Genealogy*

In his *Anxiety* seminar, Lacan reproaches Lévi-Strauss and structuralism in general for confusing "structure" with the form of the brain. "The play of structure," he writes, "of the combinatory that was so powerfully articulated in the discourse of Lévi-Strauss only rejoins for example the very structure of the brain, indeed the structure of matter, of which it would merely represent, in accordance with the form called materialist in the eighteenth-century sense, the doublet, and not even the inner lining (*doublure*)."[1] Such coincidence between symbolic structure and cerebral structure would be the sign of what Lacan calls "primary materialism."[2]

It is precisely such "primary materialism"—which is Lacan's expression of contempt for the cerebral localization of the symbolic—that I have attempted to assume and to uphold throughout my discussion. Extending the closing argument of *What Should We Do with Our Brain?* I continue to defend the thesis that the only valid philosophical path today lies in the elaboration of a new materialism that would precisely refuse to envisage the least separation, not only between the brain and thought but also between the brain and the unconscious.

It is thus such a materialism, as the basis for a *new philosophy of spirit*, that determined my definition of cerebrality as an axiological principle entirely articulated in terms of the formation and deformation of neuronal connections. The "symbolic" is obviously not far away, since the elementary form of the brain is the emotional and logical core where the processes of auto-affection constitute all identity and all history.

This process is radically exposed to the possibility of an accident that might destroy it and thereby interrupt the continuity of the psychic personality. Such vulnerability is the major question of contemporary psychopathology. The study of brain damage reveals that traumas and wounds have a new signification that psychoanalysis can only ignore at the price of failing to grasp present-day psychic suffering.

This new signification is linked to negative or destructive plasticity. Its result can be characterized as a metamorphosis unto death or as a form of death in life marked by affective indifference.

Recognizing the existence of negative plasticity beyond any promise of remission or any soteriological horizon is the necessary prelude to any attempt to account for psychic suffering today. The confrontation of the etiological regimes of sexuality and cerebrality, and thus of psychoanalysis and neurology, can be fruitful only if it begins from such recognition.

In order to establish such recognition, I have attempted to elaborate the concept of the *material event*. Conceived as an accident or threat of destruction, this material event—which goes beyond the Lacanian triad of the symbolic, the real, and the imaginary—results as much from the contingency of its occurrence as from the internal work of the drive, work that demands a new understanding of the contingency and the necessity of the death drive.

To insist on the role of destructive plasticity beyond all horizon of redemption does not amount to denying the possibilities of new therapies. This is not a matter of despair or pessimism. I am simply arguing that, before interrogating the hypothetical possibility of a "beyond of the beyond" of the death drive, before asking how to treat or how to heal, it is important, according to the most elementary logic, to inquire what those who suffer are suffering from.

This is why we have had to spend so much time profiling the new wounded. Our inquiry revolves around the identification of evil. Defining the characteristics of today's traumas—characteristics that turn out to be geopolitical—is indeed the prolegomenon to any therapeutic enterprise.

The destructive event that—whether it is of biological or sociopolitical origin—causes irreversible transformations of the emotional brain, and thus of a radical metamorphosis of identity, emerges as a constant existential possibility that threatens each of us at every moment. At every instant, we are all susceptible to becoming *new wounded*, prototypes of ourselves without any essential relation to the past of our identities. Alzheimer's disease is a particularly important example of such loss. A form of life appears that bids farewell to all the subject's old modes of being.

This type of transformation unto death, this survival without sublation, is not only visible in cases of severe brain lesions but is also the globalized form of trauma—appearing in the aftermath of wars, terrorist attacks, sexual abuse, and all types of oppression or slavery. Today's violence consists in cutting the subject away from its accumulated memories.

We have seen that, for Freud, every psychic disturbance, whether personal or collective, constitutes a regression. Herein lies the great political lesson of psychoanalysis—that the primitive is always capable of resurfacing.

> Thus the transformation of the drive, on which our susceptibility to culture is based, may also be permanently or temporarily undone by the impacts of life. The influences of war are undoubtedly among the forces that can bring about such involution.[3]

However, it remains possible that today's new wounds do not expose the indestructible basis of humanity—the famous primitive aggressivity or cruelty. It remains possible, inversely, that we are witnessing forms of pure destruction, instantiated by disaffected individuals who have themselves been formed by destruction. *The primitive might well have definitively abandoned the unconscious.* For this reason, the brain is not exposed to the same danger as the psyche such as Freud defined it. The danger consists in the always possible—and often politically organized—volatilization of the supposedly "indestructible" core of the unconscious.

It is indeed possible that these are the new facts that challenge psychoanalysis to transform itself from top to bottom; that the new wounded do not regress; that the force of trauma, whether political or lesional, never derives from lifting repression; that patients' words reveal nothing; that illness does not in itself constitute a form of truth with respect to the ancient history of the subject.

The distinction between organic traumas and political traumas becomes blurred precisely because of the type of event that gives rise to them—a brutal event, without signification, that tends to efface its intentionality in order appear as a blow inflicted upon any possible hermeneutics in general. The principal conclusion that one can draw from the existence of such events is that the wounds to which they give rise largely constitute temporal ruptures, existential improvisations that proceed from the nullification of the past.

To make this claim, once again, is not a matter of pessimism. To apprehend the new wounded as figures of the death drive which would no longer derive from sexual etiology is, on the contrary, a very fruitful point of departure for a clinic to come—a clinic that, much like neuropsychoanalysis seeks to do, would integrate the conjoined results of Freudianism and neurology.

Indeed, as certain psychoanalysts are beginning to observe, it is not only *traumatic neurosis* that has disappeared from the contemporary psychopathological landscape but also *transference neurosis*.[4] A deserted, emotionally disaffected, indifferent psyche is not or is no longer capable of transference. We live in the epoch of the end of transference. Feelings of love for the psychoanalyst or the therapist no longer mean anything for a psyche that can no longer love or hate. As Daniel Widlöcher courageously writes, we must think "beyond transference neurosis" where "the model of transference neurosis no longer applies."[5] Davoine and Gaudillière, as well, underscore that "the handling of transference is very different from its classical use in psychoanalysis."[6] From now on, it is a matter of "becoming subject to the other's suffering, especially when this other is unable to feel anything."[7]

It is clear that insisting on the brutal character of the accident or the catastrophe, on the external dimension of violence, could suggest that the subject counts for nothing in what comes to pass.[8] That the victim of trauma is totally incapable of symbolically reappropriating, in one way or another, the destructive event; that the executioners are not implicated in what they do (one thinks of the coolness of serial killers) and thus are not *responsible*—I have not made such claims because I have always attempted to show the link between external violence and the play of the drives.

Nonetheless, how could we deny that the new wounded *call responsibility into question*? Not to respond, no longer to respond, is also not to wait for or ask for a response. Hasn't this desertion of the structure of transference, which constitutes the demand for response, become incontestable? To say so is not to exculpate anyone, but to establish that the "subject supposed to know" has been deposed in patients who do not want either to know or not to know.

Taking seriously the motif of the end of transference neurosis obviously requires that we elaborate anew the role of the analyst, the meaning of demand, and, consequently, the meaning of response. The therapist must discover a way "to become the subject of the other's suffering" without thereby entering into transference with him. The coolness of patients would thus challenge psychoanalysis to rethink the *neutrality* of the analyst. The possibility of elaborating a *nontransferential clinic* ultimately appears as the major stake of the confrontation between psychoanalysis and neurology, opening neuropsychoanalysis to the promise of its future.

Between psychoanalysis and neurology, it is precisely the sense of "the other" that is displaced. To recognize in cerebrality the other of sexuality—which Freud always sought without ever finding it—is also to interrogate neurology's capacity to welcome conceptually an alterity that, in a certain sense, it found but did not seek.

To gather the other's pain is not to take his place, but to restore it to him. This is what I learned—too late, too early—from a patient with Alzheimer's.

PREAMBLE

1. Marcel Proust, *Sodom and Gomorrah*, trans. and intro. John Sturrock (New York: Penguin Books, 2002), 159–160.

2. " . . . we suddenly find ourselves able to perceive our own absence." Marcel Proust, *Guermantes' Way* (New York: Penguin Books, 2005), 134.

3. *Le livre noir de la psychanalyse: Vivre, penser, et aller mieux sans Freud*, ed. Catherine Meyer, Mikkel Borch-Jakobsen, Didier Pleux, and Jacques van Rillaer (Paris: Éditions des Arènes, 2005).

4. I am thinking, for example, of Gérard Pommier, *Comment les neurosciences démontrent la psychanalyse* (Paris: Flammarion, 2004), or of François Ansermet and Pierre Magistretti, *The Biology of Freedom: Neural Plasticity, Experience, and the Unconscious*, trans. Susan Fairfield (New York: Other Press, 2007).

5. See Catherine Malabou, *What Should We Do with Our Brain?*, trans. Sebastian Rand (New York: Fordham University Press, 2008).

6. Bruno Bettelheim, *The Empty Fortress* (New York: Free Press, 1967), 8.

7. Guillaume Apollinaire, "The New Spirit and the Poets," in *Symbolist Art Theories*, ed. Henri Dorra (Berkeley: University of California Press, 1995), 311.

INTRODUCTION

1. Sigmund Freud, *The Standard Edition of the Complete Psychological Works of Sigmund Freud*, ed. and trans. James Strachey (London: Hogarth Press, 1950), 7:279. [Hereafter abbreviated as *SE*.]

2. Olivier Postel-Vinay, "Le cerveau et l'amour," *La Recherche* 3 (November 2004): 32–39.

3. Mark Solms and Oliver Turnbull, *The Brain and the Inner World: An Introduction to the Neuroscience of Subjective Experience* (New York: Other Press, 2002), 116.

4. Jean-Didier Vincent, *Biologie des passions* (Paris: Odile Jacob, 2002), 317.

5. The term "endogenous" characterizes something that emerges within the body or within an organism, what derives from an internal cause. To the contrary, the term "exogenous" designates what intervenes from the outside and results from external causes.

6. Paul Ricoeur, *Freud and Philosophy*, trans. Denis Savage (New Haven: Yale University Press, 1970), 65–67.

7. The German word for "traumatism" is *Trauma*, which is the term that Freud employs. I will use "trauma" and "traumatism" interchangeably. [Since, in English, the word "trauma" covers both "trauma" and "traumatism," and since the author uses these terms interchangeably, I have simply translated both using the word "trauma."—Trans.]

8. Freud, *SE*, 18:12. The same affirmation is repeated later in the text (33): "a gross physical injury caused simultaneously by the trauma diminishes the chances that a neurosis will develop."

9. Freud, *SE*, 17:210.

10. Ibid., 212.

11. Ibid., 207.

12. Ibid., 208.

13. Freud, *SE*, 3:193.

14. Freud, *SE*, 7:275.

15. Ibid.

16. Freud, *SE*, 18:12.

17. Ibid.

18. Ibid.

19. [The French word *événementiel* is notoriously difficult to translate. Throughout this text, in accordance with current practice, I have primarily translated it using the serviceable but less-than-satisfying neologism "evental." Likewise, I have translated *événementialité* as "eventality." In order to avoid overuse of this word, however, I have often translated Malabou's expression *régime événementiel* simply as "regime of events."—Trans.]

20. Freud, *SE*, 4:185n.

21. Freud, *SE*, 3:274.

22. This phrase appears in "Introduction to 'The Psychoanalysis of War Neurotics,'" *SE*, 17:210.

23. Antonio Damasio, *The Feeling of What Happens* (San Diego: Harcourt, 1999), 13–15.

24. Mark Solms is author of two particularly relevant works: *A Moment of Transition: Two Neuroscientific Articles by Sigmund Freud* (with Michael Saling), (London: Karnac Books, 1990), and *The Neuropsychology of Dreams: A Clinico-Anatomical Study* (Mahwah, N.J.: Lawrence Erlbaum Associates, 1997). See also

an article recently published in French, "Psychanalyse et neurosciences," in *Pour la science*, October 2004, 77–81.

25. Neuropsychoanalysis is a strain of thought that has given rise in the United States to an international society (The International Neuropsychoanalysis Society) and the publication of a journal, *Neuropsychoanalysis*. This movement has brought together many of the best known neuroscientists in the world today, such as Antonio Damasio, Erik Kandell (Nobel Prize for Medicine, 2000), Joseph LeDoux, and Benjamin Libet, just to name a few. In France, the psychoanalysts, psychiatrists, neuropsychologists, and neurobiologists who are most interested in this concept are primarily André Green, Daniel Widlöcher (presently, the director of the Association psychanalyse et psychothérapie [APEP]), Sylvain Missonier, Jean-Pol Tassin, Nicolas Georgieff, and Marc Jeannerod.

26. At the Hôpital de la Salpêtrière in Paris, a group of neuropsychologists, psychiatrists, and psychoanalysts meets regularly at the behest of Daniel Widlöcher in order to discuss cases of patients who have been doing "psycho-analytic psychotherapies" after incurring brain lesions that resulted in aphasia, disturbances of memory, or disturbances in their emotional behavior or life skills. Nonetheless, in France, the concept of neuropsychoanalysis is often called into question. In February 2005, for example, there was a conference on neuropsychoanalysis at the Hôpital Necker entitled "The Anglo-Saxon Concept of Neuro-Psychoanalysis: Its Interest and Its Limits."

27. Solms and Turnbull, *The Brain and the Inner World*, 298; Sigmund Freud, *Beyond the Pleasure Principle*, SE, 18:7.

28. Solms, "Psychanalyse et neurosciences," 81.

29. Solms and Turnbull, *The Brain and the Inner World*, vii.

30. Ibid., vii–viii.

31. See the interesting analysis of Jean-Michel Thurin that is reproduced in the report from a roundtable discussion among Jean-Michel Thurin, Pierre-Henri Castel, and Bernard Perret entitled "Psychotherapies: Which Evaluation?" Discussing Kandell's work, the author notes that in the United States today, "the model is that of relations to the environment, but also, at the same moment, we see a re-emergence of psychoanalysis, which had been completely eclipsed for many years. . . . We can thus observe a double movement: there are just as many works concerned with the affective life and memory on the cerebral level but, at the same time, psychoanalysis has regained its civil rights." *Esprit*, "Les guerres du sujet," November 2004, 157–158.

32. Solms and Turnbull, *The Brain and the Inner World*, viii.

33. Alexander R. Luria (1902–77) is the author of numerous fundamental works such as *Traumatic Aphasia: Its Syndromes, Psychology, and Treatment*, trans. Douglas Bowden (Berlin: Mouton de Gruyter, 1970); *The Working Brain: An*

Introduction to Neuropsychology, trans. Basil Haigh (New York: Basic Books, 1973); *Higher Cortical Functions in Man*, trans. Basil Haigh (New York: Basic Books, 1980); *The Making of a Mind: A Personal Account of Soviet Psychology*, trans. Mike Cole (Cambridge, Mass.: Harvard University Press, 1979); *The Mind of a Mnemonist: A Little Book About a Vast Memory*, trans. Jerome Bruner (Cambridge, Mass.: Harvard University Press, 1987); and *The Man With a Shattered World: The History of a Brain Wound*, trans. Lynn Solotaroff (Cambridge, Mass.: Harvard University Press, 1987).

34. Solms and Turnbull, *The Brain and the Inner World*, viii.

35. Oliver Sacks, *The Man Who Mistook His Wife for a Hat* (New York: Simon & Schuster, 1998), 4.

36. Solms and Turnbull, *The Brain and the Inner World*, ix.

37. Ibid.

38. Whence the hyphen that, in French, still divides the name *neuro-psycho-analyse*.

39. According to Damasio's formulation in *The Feeling of What Happens*, 53.

40. This expression is used several different times in Chapter 3 of *The Ego and the Id* ("The Ego and the Superego"), *SE*, 19:29, 30, 34.

41. Sacks, *The Man Who Mistook His Wife for a Hat*, 36.

42. François Lebigot and Nicolas Prieto, "Les soins psychiques précoces en cas de catastrophe" (Early Psychic Care in Case of Disaster), http://bit.ly/uSGQz6.

43. "For neurobiology," Françoise Davoine and Jean-Max Gaudillière affirm, "this clinical case is well on the way to becoming as famous as the Schreber case is for the psychoanalysis of the psychoses." Françoise Davoine and Jean-Max Gaudillière, *History Beyond Trauma: Whereof One Cannot Speak, Thereof One Cannot Stay Silent*, trans. Susan Fairfield (New York: Other Press, 2004), 49–50.

44. Antonio Damasio, *Descartes' Error: Emotion, Reason, and the Human Brain* (New York: Penguin, 2005), 7.

45. Ibid.

46. Mark Solms, *The Brain and the Inner World*, 3.

47. Ibid., 4.

48. Davoine and Gaudillière, *History Beyond Trauma*, 50.

49. Ibid., 106.

50. For a definition of plasticity, see in particular my book, *What Should We Do with Our Brain?* (New York: Fordham University Press, 2008).

51. Freud, *SE*, 14:285–286.

52. Ibid., 286.

53. Ibid.

54. Damasio, *Descartes' Error*, 56.

PART I INTRODUCTION: THE "NEW MAPS" OF CAUSALITY

1. The subtitle of this chapter derives from a play on the title of Daniel Widlöcher's book *Les nouvelles cartes de la psychanalyse* (Paris: Odile Jacob, 1996). Immanuel Kant, *Critique of Pure Reason*, trans. and ed. Paul Guyer and Allen W. Wood (Cambridge: Cambridge University Press, 1997), 536.

2. Marc Jeannerod, "Neurosciences et psychiatrie. Attirance ou repulsion?" *Les Temps Modernes* 630–631 (May–June 2005): 68–82.

3. Ibid., 77.

4. Ibid., 75.

5. André Green, *La causalité psychique: Entre nature et culture* (Paris: Odile Jacob, 1995), 86ff.

6. Sigmund Freud, *The Standard Edition of the Complete Psychological Works of Sigmund Freud*, ed. and trans. James Strachey (London: Hogarth Press, 1950), 11:36 (translation slightly modified).

7. Joëlle Proust, "La psychanalyse au risque des neurosciences," in *Le livre noir de la psychanalyse*, 652.

8. Ibid., 652–653.

1. CEREBRAL AUTO-AFFECTION

1. Sigmund Freud, *The Standard Edition of the Complete Psychological Works of Sigmund Freud*, ed. and trans. James Strachey (London: Hogarth Press, 1950), 14:118 (translation slightly modified). [Hereafter abbreviated as *SE*.]

2. Ibid.

3. Freud, *SE*, 1:296–297.

4. Freud declares: "Let us recall, then, that the nervous system had two functions: the reception of stimuli from outside and the discharge of stimuli of endogenous origin. . . . We might then conjecture that it might be our systems *phi* and *psi* each of which had assumed one of these primary obligations. The system *phi* would be the group of neurons which the external stimuli reach, the system *psi* would contain the neurons which receive the endogenous stimuli" (ibid., 303).

5. Freud, *SE*, 14:120 (translation slightly modified).

6. Ibid.

7. Freud, *SE*, 2:193–194.

8. Ibid., 198.

9. Ibid., 198–199.

10. From the article "Psychical Representative," in Jean Laplanche and Jean-Baptiste Pontalis, *The Language of Psychoanalysis*, trans. Donald Nicholson-Smith (New York: Norton, 1973), 364. See also "Ideational Representative" in ibid., where "Representative" renders *Repräsentanz*, a German term of Latin

origin that should be understood as implying delegation. *Vorstellung* is a philosophical term whose traditional French equivalent is *représentation* and whose traditional English equivalent is "idea." *Vorstellungrepräsentanz* means a delegate (in this instance, a delegate of the instinct) in the domain of representation" (203–204, translation modified).

11. Freud, *SE*, 14:152 (translation modified).

12. Ibid., 177 (translation modified). In *Repression*, Freud presents these same transformations in the following manner: "Either the drive is altogether suppressed, so that no trace of it is found, or it appears as an affect which is in some way or other qualitatively colored, or is changed into anxiety. The two latter possibilities set us the task of taking into account, as a further instinctual vicissitude, the *transformation* into *affects*, and especially into *anxiety*, of the psychical energies of the *drives*" (153, translation modified).

13. Ibid., 126.

14. Michel de Certeau, *Heterologies: Discourse on the Other*, trans. Brian Massumi (Minneapolis: University of Minnesota Press, 1986), 22–23.

15. This helps to clarify certain of Freud's assertions according to which "mental activity is bound up with the brain as it is with no other organ. We have taken a step further—we do not know how much—by the discovery of the unequal importance of the different parts of the brain and their special relations to particular parts of the body and to particular mental activities. But every attempt to go on from there to discover a localization of mental processes, every endeavor to think of ideas as stored up in nerve-cells and of stimuli as travelling along nerve-fibers, has miscarried completely. The same fate would await any theory which attempted to recognize, let us say, the anatomical position of the system *Cs.*—conscious mental activity—as being the cortex, and to localize unconscious processes in the subcortical parts of the brain. There is a hiatus here which at present cannot be filled, nor is it one of the tasks of psychology to fill it. Our psychical topography has *for the present* nothing to do with anatomy; it has reference not to anatomical localities, but to regions in the mental apparatus, wherever they may be situated in the body" (Freud, *SE*, 14:174–175).

16. In Chapter 7, I address the relationship between sexuality and division, or separation, in greater detail.

17. This neologism is formed on the model of Jacques Derrida's famous neologisms *différance*, *restance*, *destinérrance*, and so on. The undecidable grammar of such terms (both past participle and gerund) makes it possible to articulate a process that that is at once passive and active, past and progressive, settled and ongoing.

18. Jacques Lacan, *The Seminar of Jacques Lacan, Book II: The Ego in Freud's Theory and in the Technique of Psychoanalysis*, 1954–55, trans. Sylvana Tomaselli

(New York: Norton, 1991), 222. In the same text, Lacan also declares: "The notion of libido is thus a form of unification of the field of psychoanalytic effects."

19. Ibid.

20. Ibid.

21. Antonio Damasio, *The Feeling of What Happens* (San Diego: Harcourt, 1999), 40.

22. Freud, *SE*, 14:120–121.

23. Damasio, *The Feeling of What Happens*, 76.

24. Antonio Damasio, *Looking for Spinoza: Joy, Sorrow, and the Feeling Brain* (Orlando: Harcourt Books, 2003), 174.

25. Ibid., 59.

26. Damasio, *The Feeling of What Happens*, 169.

27. Ibid., 170.

28. Ibid., 154.

29. Marcel Gauchet, *L'inconscient cérébral* (Paris: Seuil, 1992), 37.

30. This would be the appropriate moment to indicate the existence of the remarkable neurons called "mirror neurons," which, as they perceive the other executing a given activity, code and activate the movement just as if it were the subject executing it himself. See Marc Jeannerod, *Le cerveau intime* (Paris: Odile Jacob, 2002), 189.

31. They make this claim against Freud's own explicit statements: "We know two kinds of things about what we call our psyche (or mental life): firstly, its bodily organ and scene of action, the brain (or nervous system) and, on the other hand, our acts of consciousness, which are immediate data and cannot be further explained by any sort of description. Everything that lies between is unknown to us, and the data do not include any direct relation between these two terminal points to our knowledge. If it existed, it would at the most afford an exact localization of the processes of consciousness and would give us no help towards understanding them" (*SE*, 23:144–145).

32. Mark Solms and Oliver Turnbull, *The Brain and the Inner World: An Introduction to the Neuroscience of Subjective Experience* (New York: Other Press, 2002), 110.

33. Freud, *SE*, 19:26.

34. Ibid.

35. Damasio, *The Feeling of What Happens*, 189.

36. Ibid., 154.

37. Ibid., 191.

38. Freud, *SE*, 14:187.

39. Ibid., 289.

40. Damasio, *The Feeling of What Happens*, 144–145.

2.   BRAIN WOUNDS: FROM THE NEUROLOGICAL NOVEL TO THE THEATER OF ABSENCE

1. Joseph LeDoux, *The Synaptic Self* (New York: Penguin Books, 2002), 304–305.

2. Antonio Damasio, *The Feeling of What Happens* (San Diego: Harcourt, 1999), 40–41.

3. Ibid., 41.

4. Ibid., 85–86.

5. Antonio Damasio, *Descartes' Error: Emotion, Reason, and the Human Brain* (New York: Penguin, 2005), xvi.

6. Antonio Damasio, *Looking for Spinoza: Joy, Sorrow, and the Feeling Brain* (Orlando: Harcourt Books, 2003), 60.

7. Damasio, *The Feeling of What Happens*, 43.

8. Damasio, *Descartes' Error*, 34–35.

9. Ibid., 36.

10. Damasio, *The Feeling of What Happens*, 96.

11. Ibid., 98.

12. Ibid., 97.

13. Ibid., 101–102.

14. Ibid., 103. On the important difference between this state of absence and locked-in syndrome, see also 292–294.

15. Damasio, *Descartes' Error*, 72.

16. Ibid., 73.

17. Damasio, *The Feeling of What Happens*, 162–167.

18. Damasio, *Descartes' Error*, 64.

19. Oliver Sacks, *The Man Who Mistook His Wife for a Hat* (New York: Simon & Schuster, 1998), viii.

20. Ibid.

21. Alexander Luria, *L'homme dont le monde volait en éclats*, trans. Fabienne Mariengof and Nina Rausch de Traubenberg (Paris: Seuil, 1995), 25.

22. A. R. Luria, *The Man with a Shattered World*, trans. Lynn Solotaroff (Cambridge, Mass.: Harvard University Press, 1972), vii.

23. Sacks, *The Man Who Mistook His Wife for a Hat*, ix.

24. Luria, *The Man with a Shattered World*, 100 (translation significantly modified).

25. Damasio, *The Feeling of What Happens*, 90–92.

26. Ibid., 90.

27. Samuel Beckett, *Happy Days* (New York: Grove Press, 1961), 21.

28. Gilles Deleuze, *Essays Critical and Clinical*, trans. Daniel Smith and Michael A. Greco (London: Verso Books, 1998), 152.

29. Ibid., 153.

30. Ibid.

3. IDENTITY WITHOUT PRECEDENT

1. Sigmund Freud, *The Standard Edition of the Complete Psychological Works of Sigmund Freud*, ed. and trans. James Strachey (London: Hogarth Press, 1950), 14:285. [Hereafter abbreviated as *SE*.].

2. Ibid., 285–286.

3. Ibid., 286.

4. Antonio Damasio, *The Feeling of What Happens* (San Diego: Harcourt, 1999), 104.

5. For example: "important neurological signs, which usually disappear in the course of childhood development, will reappear in the patient. To mention only the best known example, this happens with the grip reflex (whereby a baby automatically grasps anything placed in its hand)." Olivier Blond, "Une retombée en enfance?" in *Alzheimer, cerveau sans mémoire*, a special issue of *La Recherche*, 10 (January–March 2003): 80.

6. Cited in ibid., 80.

7. John Bayley, *Iris: A Memoir of Iris Murdoch* (London: Abacus, 1998), 243.

8. Larry Squire and Eric Kandell, *Memory: From Mind to Molecules* (New York: Henry Holt, 1999), Chap. 5.

9. These circuits are essential to homeostatic equilibrium, metabolic equilibrium, the search for food and shelter, the avoidance of predators, and reproduction.

10. Antonio Damasio, *Descartes' Error: Emotion, Reason, and the Human Brain* (New York: Penguin, 2005), 110.

11. Ibid., 111.

12. Ibid., 112.

13. Joseph LeDoux, *The Synaptic Self* (New York: Penguin Books, 2002), 270.

14. Ibid., 268. Until recently, schizophrenic disturbances were understood to result from an excessively elevated quantity of a certain neurotransmitter: dopamine. "By the mid-1970s," LeDoux continues, "it seemed that depression was due to too little monoamine transmission, just as schizophrenia was due to too much. This conclusion implied a rather simple, in fact simplistic, picture of the illness" (ibid., 274).

15. Cf. Nicolas Andreasen, "Defining the Phenotype of Schizophrenia: Cognitive Dysmetria and Its Neuronal Mechanisms," *Biological Psychiatry* 46, no. 7 (1999): 908–920.

16. LeDoux, *The Synaptic Self*, 323–324.

17. Ibid., 307, 324.

18. Freud, *SE*, 14:286.

19. Mark Solms and Oliver Turnbull, *The Brain and the Inner World: An Introduction to the Neuroscience of Subjective Experience* (New York: Other Press, 2002), 209.

20. Ibid.

21. Ibid., 209–10.

22. Françoise Davoine and Jean-Max Gaudillière, *History Beyond Trauma: Whereof One Cannot Speak, Thereof One Cannot Stay Silent*, trans. Susan Fairfield (New York: Other Press, 2004), xxi.

23. Freud, *SE*, 21:71.

4. PSYCHOANALYTIC OBJECTION: CAN THERE BE DESTRUCTION WITHOUT A DRIVE OF DESTRUCTION?

1. Oliver Sacks, *The Man Who Mistook His Wife for a Hat* (New York: Simon & Schuster, 1998), 35.

2. Ibid., 36.

3. Ibid., 37.

4. Ibid.

5. Ibid., 35.

6. Ibid., 38.

7. Ibid., 39.

8. Boris Cyrulnik, *Un merveilleux malheur* (Paris: Odile Jacob, 2002), 164.

9. Sigmund Freud, *The Standard Edition of the Complete Psychological Works of Sigmund Freud*, ed. and trans. James Strachey (London: Hogarth Press, 1950), 14:174. [Hereafter abbreviated as *SE*.].

10. Freud, *SE*, 23:229.

11. Freud, *SE*, 19:40.

12. Freud, *SE*, 18:259.

13. Freud, *SE*, 23:150.

14. Jacques Lacan, *The Seminar of Jacques Lacan, Book II: The Ego in Freud's Theory and in the Technique of Psychoanalysis*, 1954–55, trans. Sylvana Tomaselli (New York: Norton, 1991), 37. In J. Laplanche and J.-B. Pontalis, *The Language of Psychoanalysis*, trans. David Nicholson-Smith (New York: Norton, 1974), the authors recall that many French translators, in order to render *Todestrieb*, initially opted for "*instinct de mort*" (death instinct) rather than "*pulsion de mort*" (death drive).

15. It should be noted that Freud never uses this word. Instead, he speaks of a "principle of constancy" or a "principle of inertia."

16. Lacan, *The Ego in Freud's Theory*, 80.

17. "The term 'Nirvana,' which was given currency in the West by Schopenhauer, is drawn from Buddhism, where it connotes the 'extinction' of human desire, the abolition of individuality when it is fused in the collective soul, a state of quietude and bliss." Laplanche and Pontalis, *The Language of Psychoanalysis*, 272.

18. Freud, *SE*, 18:55–56.

19. In "The Economic Problem of Masochism," Freud insists upon the distinction between the two principles: "We must perceive that the Nirvana principle, belonging as it does to the death drive, has undergone a modification in living organisms through which it has become the pleasure principle; and we shall henceforward avoid regarding two principles as one" (Freud, *SE*, 19:160). But the important point is that, whether they are identical or different, or identical and different, the principle of constancy and the Nirvana principle both aim to lessen the quantum of excitation, to liquidate tension, and to postpone unpleasure.

20. Freud, *SE*, 13:38.

PART II INTRODUCTION: FREUD AND PREEXISTING FAULT LINES

1. Sigmund Freud, *The Standard Edition of the Complete Psychological Works of Sigmund Freud*, ed. and trans. James Strachey (London: Hogarth Press, 1950), 22:59. [Hereafter abbreviated as *SE*.]

2. Freud, *SE*, 18:12.

3. Freud, *SE*, 16:275.

4. Freud, *SE*, 18:29.

5. Freud, *SE*, 17:210.

6. Freud, *SE*, 18:12.

7. Herman Oppenheim, *Die traumatischen Neurosen* (Berlin: August Hirschwald, 1892).

8. The term "war neurosis" was coined by the German psychiatrist Honigman in 1907.

9. Freud, *SE*, 18:12.

10. Ibid., 33.

11. "Introduction to *Psychoanalysis and the War Neuroses*" is the text that Freud included as a preamble to the published version of the conference proceedings. Participating in this congress, which took place on the premises of the Hungarian Academy of Sciences, were representatives of the Budapest mayoralty, representatives of the governments of Hungary, Austria, and Germany, high-ranking military doctors, and psychoanalysts. The proceedings collect the contributions of Ferenczi, Abraham, and Simmel, along with a text by Jones added after the fact.

12. Louis Crocq, *Les traumatismes psychiques de guerre* (Paris: Odile Jacob, 1999), 254. In order to argue this point, Crocq refers to "Introduction to *Psychoanalysis and the War Neuroses*."

13. Karl Abraham et al., *Psychoanalysis and the War Neuroses* (London: International Psychoanalytic Press, 1922), 13, 17. As Crocq recalls, "Ferenczi,

as the head physician of a cavalry battalion near Vienna (which allowed him to pursue his psychoanalysis with Freud), and then, from 1916 onward, as a neuropsychiatrist at the Maria Valeria Hospital in Budapest, had the chance to examine many patients afflicted with war neuroses" (*Les traumatismes psychiques de guerre*, 248). It was on September 28, 1918, that he presented his report to the Budapest Congress.

14. Kurt R. Eissler, *Freud as an Expert Witness: The Discussion of War Neuroses between Freud and Wagner-Jauregg*, trans. Christine Trollope (Madison, Conn.: International Universities Press, 1986). Original title: *Freud und Wagner-Jauregg vor der Kommission zur Erhebung militärischer Pflichtverletzungen* (Vienna: Löcker Verlag, 1979).

15. Back cover of the French edition, Kurt R. Eissler, *Freud sur le front des névroses de guerre*, trans. Madeleine Drouin, Anne Porge, Erik Porge, and Anne-Marie Vindras (Paris: Presses Universitaires de France, 1992).

16. Freud, *SE*, 17:211–215.

17. Nonetheless, Freud exculpated him of any "intention to harm."

18. Erik Porge, "Introduction" to *Freud sur le front des névroses de guerre*, vii.

5.   WHAT IS A PSYCHIC EVENT?

1. Sigmund Freud, *The Standard Edition of the Complete Psychological Works of Sigmund Freud*, ed. and trans. James Strachey (London: Hogarth Press, 1950), 2:6. [Hereafter abbreviated as *SE*.]

2. Freud, *SE*, 7:276. "Thus it was no longer a question of what sexual experiences an individual had in his childhood, but rather of his reaction to those experiences," 276–277.

3. Freud, *SE*, 6:191.

4. Ibid.

5. Ibid., 256–257.

6. Ibid., 257.

7. Ibid.

8. Freud, *SE*, 2:xxxi. Later in the same text, Freud writes: "It would be unfair if I were to try to lay too much of the responsibility for this development upon my honored friend Dr. Josef Breuer," 256.

9. Ibid., xxix.

10. Ibid., 244.

11. Ibid., 5–6.

12. Ibid., 5.

13. Ibid., 4.

14. Ibid., 6.

15. Ibid.

16. Ibid., 192.

17. Ibid., 193–194.

18. Ibid., 203.

19. Ibid., 204.

20. Ibid., 201.

21. Ibid., 209.

22. Ibid., 257.

23. See, on this point, *Beyond the Pleasure Principle*: "Such an event as an external trauma is bound to provoke a disturbance on a large scale in the functioning of the organism's energy and to set in motion every possible defensive measure" (Freud, *SE*, 18:29).

24. "This compound, as we know, was itself merely soldered together" (Freud, *SE*, 9:161).

25. Ibid., 229.

26. Jacques Lacan, *The Seminar of Jacques Lacan, Book VII: The Ethics of Psychoanalysis*, ed. Jacques-Alain Miller, trans. Dennis Porter (New York: Norton, 1992), 139.

27. Freud, *SE*, 3:195, 215.

28. Ibid., 199.

29. Freud, *SE*, 2:290–291.

30. Freud, *SE*, 3:191.

31. Ibid., 144.

32. Ibid., 148–149.

33. Ibid., 149.

34. Ibid., 207.

35. Ibid., 207–208.

36. Ibid., 200.

37. Ibid., 217.

38. Ibid., 203.

39. Freud, *SE*, 14:17–18.

40. Freud, *SE*, 5:620.

41. See J. Laplanche and J.-B. Pontalis, *The Language of Psychoanalysis*, trans. David Nicholson-Smith (New York: Norton, 1974), 407: "He does not see seduction, essentially, as a concrete fact which can be assigned its place in the subject's history; instead, he looks upon it as a structural datum whose only possible transposition into historical terms would be in the form of a myth."

42. Freud, *SE*, 9:229.

43. Freud, *SE*, 4:41–42.

44. Freud, *SE*, 1:162.

45. This mimetic faculty is also discussed in "Sexuality in the Etiology of the Neuroses," where Freud speaks of the way in which hysteria "imitates so many organic affections," *SE*, 3:270.

46. Freud, *SE*, 1:169.

47. Ibid., 164.

48. See the example of the "completely inert" arm, ibid., 164.

49. Freud, *SE*, 7:274.

50. Ibid., 279.

51. The epistemological signification of sexuality, it is worth recalling, is first and last of all, for Freud, an "etiological signification." See in particular the article "The Etiological Significance of Sexual Life" in *Two Encyclopedia Articles* (1923), *SE*, 18:243.

52. Freud, *SE*, 23:186.

6. THE "LIBIDO THEORY" AND THE OTHERNESS OF THE SEXUAL TO ITSELF:

TRAUMATIC NEUROSIS AND WAR NEUROSIS IN QUESTION

1. Sexuality, Freud writes, "is to be understood in the extended sense in which it is used in psychoanalysis and is not to be confused with the narrower concept of 'genitality'" [Sigmund Freud, *The Standard Edition of the Complete Psychological Works of Sigmund Freud*, ed. and trans. James Strachey (London: Hogarth Press, 1950], 17:208. [Hereafter abbreviated as *SE*.]).

2. Freud, *SE*, 14:125.

3. Freud, *SE*, 17:137.

4. Freud, *SE*, 18:51–52. In this passage, Freud is obviously referring to his 1914 text "On Narcissism: An Introduction."

5. Freud, *SE*, 17:137.

6. See Chapter 1.

7. This formulation appears in Freud's "On Narcissism: An Introduction" (*SE*, 14:73).

8. See ibid., 79–80.

9. Freud, *SE*, 18:53.

10. Freud, *SE*, 19:255–256.

11. This summary makes it possible to understand a note from *The Interpretation of Dreams* devoted to the accusation of pansexualism leveled against psychoanalysis, the claim that "all dreams have a sexual content." Freud denounces "the unconscientious manner in which critics are accustomed to perform their functions, and the readiness with which my opponents overlook the clearest statements. . . . The situation would be different if 'sexual' was being used by my critics in the sense in which it is now being employed in psychoanalysis—in the sense of 'eros.' But my opponents are scarcely likely to have in mind the interesting problem of whether all dreams are created by 'libidinal' instinctual forces as contrasted with 'destructive' ones" (Freud, *SE*, 4:160–161).

12 Freud, *SE*, 7:167ff.

13. Ibid., 183.

14. Ibid., 169.

15. Ibid., 183–184.

16. Ibid., 183.

17. Freud, *SE*, 23:151.

18. Freud, *SE*, 14:84.

19. Freud, *SE*, 7:204.

20. Freud, *SE*, 14:84.

21. Ibid., 76.

22. Ibid., 82.

23. Freud, *SE*, 12:70.

24. Freud, *SE*, 14:82.

25. Freud, *SE*, 17:208–209.

26. Ibid., 209.

27. Ibid., 211.

28. It should be recalled that Freud defines transference neurosis as an object neurosis—that is, as a neurosis that bears upon a sexual object, even if this object, in war neuroses, is an internal object, a modification of the ego provoked by the internalization of an enemy body that becomes a danger. See ibid., 210.

29. Ibid., 209.

30. Ibid., 210.

31. Ibid., 209.

32. Freud, *SE*, 18:12.

33. Freud, *SE*, 17:211.

34. Ibid., 212.

35. Ibid.

36. Ibid., 209.

37. Freud, *SE*, 18:33.

38. Freud, *SE*, 17:209.

39. Freud, *SE*, 14:82.

40. Ibid., 83.

41. Freud, *SE*, 18:12.

42. Ibid., 33.

43. Ibid.

44. Schreber withdrew "his libido from the people and things in the external world without finding substitutes for them in his fantasy. When this substitution does take place subsequently, it seems to be secondary, part of the recovery process that would return the libido to its object." Strangely, narcissistic withdrawal would thus be the prelude to new attachments. "From this

feature [of dementia praecox] we infer that the repression is effected by means of detachment of the libido. Here once more we may regard the phase of violent hallucinations as a struggle between repression and an attempt at recovery by bringing the libido back again on to its objects." This is why Freud can say that Schreber's psychosis had a "relatively favorable outcome." "The delusional formation, which we take to be the pathological product, is in reality an attempt at recovery, a process of reconstruction." Schreber is pursuing "something approximating to a recovery" (Freud, *SE*, 12:77).

45. See, in particular, "Two Encyclopedia Articles," *SE*, 18:258–259.

46. See *Beyond the Pleasure Principle*, in particular the first paragraphs of Chapter 6; *SE*, 18:44.

47. Ibid., 36.

48. Ibid., 38.

49. Ibid.

50. Ibid.

51. "It was still an easy matter at that time for a living substance to die; the course of its life was probably only a brief one, whose direction was determined by the chemical structure of the young life. For a long time, perhaps, living substance was being constantly created afresh and easily dying, till decisive external influences altered in such a way as to oblige the still surviving substance to diverge ever more widely from its original course of life and to make ever more complicated detours before reach its aim of death. These circuitous paths to death, faithfully kept to by the conservative drives, would thus present us today with the picture of the phenomena of life. If we firmly maintain the exclusively conservative nature of the drives, we cannot arrive at any other notions as to the origin and aim of life" (Ibid., 38–39).

52. Ibid., 38.

53. Ibid., 40–41.

54. Ibid., 39.

55. Ibid.

56. In *An Outline of Psychoanalysis*, Freud writes: "Thus it may in general be suspected that an *individual* dies of his internal conflicts." *SE*, 23:150.

57. Gilles Deleuze, *Difference and Repetition*, trans. Paul Patton (New York: Columbia University Press, 1994), 259.

58. Freud, *SE*, 18:258.

59. Freud, *SE*, 22:209.

60. Freud, *SE*, 19:41.

61. Freud, *SE*, 22:211.

62. Freud, *SE*, 19:41–42. In addition, on the basis of this demonstration, Freud elaborates the movement of *regression* from one sexual stage back to another.

7. SEPARATION, DEATH, THE THING, FREUD, LACAN, AND THE MISSED ENCOUNTER

1. See the concluding developments of Chapter 1.

2. Gilles Deleuze, *Difference and Repetition*, trans. Paul Patton (New York: Columbia University Press, 1994), 114.

3. Sigmund Freud, *The Standard Edition of the Complete Psychological Works of Sigmund Freud*, ed. and trans. James Strachey (London: Hogarth Press, 1950), 19:58. [Hereafter abbreviated as *SE*.]

4. Ibid.

5. Freud, *SE*, 22:86.

6. Ibid. [Throughout this section, I have modified Strachey's translation to accord with the French translation of Freud and with Malabou's argumentation —Trans.]

7. Ibid.

8. Ibid., 89.

9. Ibid., 93.

10. Ibid., 84.

11. Ibid., 86.

12. Ibid., 87.

13. Ibid., 87–88.

14. Ibid., 88.

15. Freud, *SE*, 20:135.

16. Jacques Lacan, *Le Séminaire, tome X: L'angoisse* (Paris: Seuil, 2004), 108.

17. Freud, *SE*, 20:139–140.

18. Ibid., 125–126.

19. Freud, *SE*, 22:89–90.

20. Freud, *SE*, 19:58.

21. Freud, *SE*, 20:129–130.

22. Ibid.

23. Ibid.

24. Ibid.

25. Freud, *SE*, 17:225.

26. Freud, *SE*, 19:58.

27. Freud, *SE*, 17:235. On the relation between the primitive, the double, and the representation of death, see also the second part of "Thoughts for the Times on War and Death," where Freud establishes the genesis of the relation between the "civilized" man to death by starting from the mentality of primitive man and studying the evolution of this relation. For further elaboration of this point, see my article, "La naissance de la mort: Hegel et Freud en guerre?" in *Autour de Hegel: Hommage à Bernard Bourgeois*, ed. François Dagognet and Pierre Osmo (Paris: Vrin, 2000), 319–331.

28. Ibid., 242.

29. Freud, *SE*, 20:129.

30. Ibid.

31. Freud, *SE*, 17:235.

32. Freud, *SE*, 22:81.

33. Sigmund Freud, *A Phylogenetic Fantasy: Overview of the Transference Neuroses*, ed. Ilse Grubich-Simitis, trans. Axel Hoffer and Peter T. Hoffer (Cambridge, Mass.: Harvard University Press, 1987), 13–14. [Once again, I have rendered *Realangst* as "real anxiety" rather than "realistic anxiety" in order to accord with Malabou's own translation of the German term—Trans.]

34. Ibid., 14. Here is the entire passage: "We have carried on a long dispute over whether real anxiety or anxiety of longing is the earlier of the two; whether the child changes his libido into real anxiety because he regards it as too great, dangerous, and thus arrives at an idea of danger, or whether he rather yields to a general anxiousness and learns from this also to be afraid of his unsatisfied libido. We were inclined to assume the former. . . . Now phylogenetic consideration seems to settle this dispute in favor of real anxiety and permits us to assume that a portion of the children bring along the anxiousness of the beginning of the Ice Age and are now induced by it to treat the unsatisfied libido as an external danger."

35. Freud, *SE*, 20:166.

36. Ibid., 168.

37. Jacques Lacan, *The Seminar of Jacques Lacan, Book VII: The Ethics of Psychoanalysis*, ed. Jacques-Alain Miller, trans. Dennis Porter (New York: Norton, 1992), 191.

38. Lacan, *L'angoisse*, 91.

39. Jacques Lacan, *The Seminar of Jacques Lacan, Book II: The Ego in Freud's Theory and in the Technique of Psychoanalysis*, 1954–55, trans. Sylvana Tomaselli (New York: Norton, 1991), 324.

40. Lacan, *The Ethics of Psychoanalysis*, 21.

41. Lacan, *The Ego in Freud's Theory*, 313. "There is only absence if you suggest that there may be a presence where there isn't one."

42. Jacques Lacan, *Écrits: The First Complete Edition in English*, trans. Bruce Fink in collaboration with Héloïse Fink and Russell Grigg (New York: Norton, 2006), 17.

43. Ibid., 324.

44. Lacan, *The Ethics of Psychoanalysis*, 71.

45. Ibid., 129–130.

46. Ibid., 118.

47. Freud introduces the dream at the very beginning of Chapter 7, *SE*, 5:509.

48. Jacques Lacan, *The Seminar of Jacques Lacan, Book XI: The Four Fundamental Concepts of Psychoanalysis*, trans. Alan Sheridan (New York: Norton, 1977), 53–54.

49. Ibid., 54.

50. Ibid.

51. Ibid., 55.

52. Aristotle, *Aristotle's Physics I, II*, trans. and intro. W. Charlton (Oxford: Oxford University Press, 1970), 33.

53. See Catherine Malabou, *The Future of Hegel: Plasticity, Temporality, Dialectic*, trans. Lisbeth During (New York: Routledge, 2005), 158–162.

54. Lacan, *The Four Fundamental Concepts of Psychoanalysis*, 53.

55. Ibid., 55.

56. Freud, *SE*, 5:509.

57. Lacan, *The Four Fundamental Concepts of Psychoanalysis*, 58.

58. Ibid.

59. Ibid., 59.

60. Ibid., 58.

61. Ibid.

62. Ibid., 59.

63. Ibid., 62 [translation slightly modified—Trans.].

64. Ibid. [translation slightly modified—Trans.].

65. Ibid., 64.

66. Lacan, *L'angoisse*, 162; see also 78.

8.   NEUROLOGICAL OBJECTION: REHABILITATING THE EVENT

1. Kurt R. Eissler, *Freud as an Expert Witness: The Discussion of War Neuroses between Freud and Wagner-Jauregg*, trans. Christine Trollope (Madison, Conn.: International Universities Press, 1986), 162.

2. Ibid., 205. See also Appendix 7, William Erb's report on Kauder's illness, 403–405.

3. Ibid., 219.

4. Ibid., 225–226.

5. J. Laplanche and J.-B. Pontalis, *The Language of Psychoanalysis*, trans. David Nicholson-Smith (New York: Norton, 1974), 472.

6. Sigmund Freud, *The Standard Edition of the Complete Psychological Works of Sigmund Freud*, ed. and trans. James Strachey (London: Hogarth Press, 1950), 18:14 (translation modified). [Hereafter abbreviated as *SE*.]

7. Freud, *SE*, 17:207.

8. Freud, *SE*, 22:93.

9. Erik Porge, introduction to Kurt R. Eissler, *Freud sur le front des névroses de guerre*, trans. Madeleine Drouin, Anne Porge, Erik Porge, and Anne-Marie

Vindras (Paris: Presses Universitaires de France, 1992), xv. Freud's students in question are Abraham, Pfister, Jones, and Ferenczi.

10. Ibid., xv–xvi.

11. Ibid., xvii.

12. Ibid., xvii–xviii.

13. Louis Crocq, *Les traumatismes psychiques de la guerre* (Paris: Odile Jacob, 1999), 257.

14. Ibid., 130.

15. This work gave rise, in 1939, to the publication of *The Individual and his Society* (New York: Columbia University Press, 1939), written in collaboration with Cora du Bois.

16. Abram Kardiner and Herbert Spiegel, *War Stress and Neurotic Illness* (New York: P. B. Hoeber, 1947). The 1941 work treats cases of war neuroses from the First World War; the revised work treats cases of wounded men from the Second World War.

17. Ibid., 286.

18. Ibid.

19. Ibid.

20. Françoise Davoine and Jean-Max Gaudillière, *History Beyond Trauma: Whereof One Cannot Speak, Thereof One Cannot Stay Silent*, trans. Susan Fairfield (New York: Other Press, 2004), 112.

21. T. W. Salmon, *The Care and Treatment of Mental Diseases and War Neuroses (Shell Shock) in the British Army* (New York: War Work Committee of the National Committee for Mental Hygiene, 1917).

22. Cf. François Lebigot and Nicolas Prieto, *Les soins psychiques précoces en cas de catastrophe* (http://bit.ly/uSGQz6), 11–13. The example discussed in this article is that of a medico-psychological cell deployed during a *plan rouge* after a gas explosion in the vicinity of Lyon. "On April 5, 2001, just before six o'clock in the evening, a gas leak in a building in Villeurbanne caused a violent explosion that killed two people and wounded nine" (2).

23. Ruth Leys, *Trauma: A Genealogy* (Chicago: University of Chicago Press, 2000), 2.

24. Crocq, *Les traumatismes psychiques de guerre*, 22.

25. Ibid.

26. See the list of traumatizing factors undergone by the victims of the war in Kosovo provided by Louis Crocq. Ibid., 196.

27. Judith Lewis Herman, *Trauma and Recovery: The Aftermath of Violence— from Domestic Abuse to Political Terror* (New York: Basic Books, 1992), 33–34.

28. Crocq, *Les traumatismes psychiques de guerre*, 273.

29. Ibid.

30. Ibid., 272. It is interesting to note that traumatic illness is also distinct from *psychosis*. Traumatic illness responds to the intrusion of a "brute real" that

cannot be incorporated into a fantasy, "while psychosis, through the ruse of delusion or delusional structure, allows for a certain acceptance of the real" (152–153).

31. Ibid., 34.

32. Ibid., 137.

33. Ibid. My emphasis.

34. Ernst Simmel, *Kriegs-Neurosen und Psychisches Trauma. Ihre gegenseitigen Beziehungen dargestellt auf Grund psycho-analytischer, hypnotischer Studien* (Munich and Leipzig: Otto Nemnich, 1918), cited in Crocq, 138; see also 269–270.

35. Crocq, *Les traumatismes psychiques de guerre*, 270. My emphasis.

36. Ibid. On this point, see also Herman, *Trauma and Recovery*, 29: "Combat leaves a lasting impression on men's minds, changing them as radically as any crucial experience through which they live."

37. Otto Fenichel, *The Psychoanalytic Theory of the Neurosis* (New York: Kegan Paul, Trench, Trubner & Co., 1946), 502.

38. Laplanche and Pontalis, *The Language of Psychoanalysis*, 471–472.

39. Crocq, *Les traumatismes psychiques de guerre*, 206.

40. American Psychiatric Association, DSM-III-R, 1987, third revised edition of the *Diagnostic and Statistical Manual of Mental Disorders* (Washington, D.C.: American Psychiatric Association, 1987), 247.

41. Jean-Bernard Andro and Claude Barrois, *L'effroi des hommes* (Paris: FR3, 1991). See also Claude Barrois, *Les névroses traumatiques* (Paris: Dunod, 1988), and "Souvenir de l'enfer et enfer de souvenir," in *Stress, psychiatrie et guerre*, Symposium de l'Association Mondiale de Psychiatrie, Paris, June 26–27 (Paris: Servier, 2003), 143–149. Barrois was the chief physician at the army teaching hospital in Val-de-Grâce.

42. Davoine and Gaudillière, *History Beyond Trauma*, 114.

43. Ibid., 114–115.

44. Herman, *Trauma and Recovery*, 28: "Not until the women's liberation movement of the 1970s was it recognized that the most common post-traumatic disorders are those not of men in war but of women in civilian life."

45. Ibid., 119.

46. "Traumatic experience . . . is rather an experience of non-sense," Crocq, *Les traumatismes psychiques de guerre*, 275.

47. Herman, *Trauma and Recovery*, 237: "The massive communal atrocities committed during the course of wars in Europe, Asia, and Africa have focused international attention on the devastating impact of violence and have fostered the recognition that psychological trauma is indeed a worldwide phenomenon."

48. Ibid., 87.

49. Ibid., 43. On the clinical description of PTSD, see also Bessel A. van der Kolk, "The Body That Keeps Score: Memory and the Evolving Psychobiology

of Post-Traumatic Stress," *Harvard Review of Psychiatry* 1 (1994): 253; and Bessel A. van der Kolk, Alexander C. McFarlane, and Lars Weisath, *Traumatic Stress: The Effects of Overwhelming Experience on Mind, Body, and Society* (New York: Guilford Press, 1996).

50. Boris Cyrulnik, *Un merveilleux malheur* (Paris: Odile Jacob, 2002), 163. "Three chemicals—cortisol, corticoid receptors, and the secretion of CRF (Cortico-Releasing-Factor) reveal that the children of traumatized people are chronically depressed." They present the same symptoms as the traumatized themselves. "The bath of anxiety in which they develop, by constantly stimulating their emotions, ends up exhausting them. . . . Emotional adaptation to trauma is not a transitory reaction, it is an acquired mode of biological reaction" (162–163).

51. Ibid., 163.

52. Allan N. Schore, "Dysregulation of the Right Brain: A Fundamental Mechanism of Traumatic Attachment and the Psychopathogenesis of Post-traumatic Stress Disorder," *Australia and New Zealand Journal of Psychiatry* 36, no. 1 (February 2002): 20.

53. Herman, *Trauma and Recovery*, 238: "Some of the most exciting recent advances in the field derive from highly technical laboratory studies of the biologic aspects of PTSD. It has become clear that traumatic exposure can produce lasting alterations in the endocrine, autonomic, and central nervous systems."

54. Antonio Damasio, *The Feeling of What Happens* (San Diego: Harcourt, 1999), 5.

55. Ibid., 6.

56. "Disaffiliation" is a concept created by the sociologist Robert Castel in *Métamorphoses de la question sociale* (Paris: Gallimard, 1995). On this point, see also my book *What Should We Do With Our Brain?*, trans. Sebastian Rand (New York: Fordham University Press, 2008), Chapter 2, "The Central Power in Crisis."

57. Antonio Damasio, *Descartes' Error: Emotion, Reason, and the Human Brain* (New York: Penguin, 2005), 178.

58. Ibid., 19.

59. Davoine and Gaudillière, *History Beyond Trauma*, xxvi.

60. Ibid., xxv.

61. Damasio, *Descartes' Error*, 178.

62. [In English, "to touch" can also mean to affect or to injure, to compromise or to taint. In everyday language, this sense of the verb most often appears negatively when, for example, one speaks of a resource, landscape, or body remaining "untouched"—Trans.]

63. Davoine and Gaudillière, *History Beyond Trauma*, 65.

64. Ibid., 47.
65. Ibid., 51.

PART III INTRODUCTION: REMISSION AT THE RISK OF FORGETTING THE WORST

1. Mark Solms and Oliver Turnbull, *The Brain and the Inner World: An Introduction to the Neuroscience of Subjective Experience* (New York: Other Press, 2002), 278–279.

2. On the effect of such therapies upon the amygdala region, see Joseph LeDoux, *The Synaptic Self* (New York: Penguin Books, 2002), 361ff.

3. On the transformation of synaptic connections during the psychoanalytic cure, see Alain Prochiantz, *La construction du cerveau* (Paris: Hachette, 1989), 60ff.

4. Lisa Ouss, "Une clinique neuropsychoanalytique: Quels modèles," confer- ence at the Hôpital Necker, "Le concept anglo-saxon de neuro-psycho-analyse: Intérêts et limites," February 10, 2005.

5. Jacques Derrida, *Without Alibi*, ed. and trans. Peggy Kamuf (Stanford: Stanford University Press, 2002), 240.

6. Ibid.

7. Ibid.

9.   THE EQUIVOCITY OF REPARATION: FROM ELASTICITY TO RESILIENCE

1. Sigmund Freud, *The Standard Edition of the Complete Psychological Works of Sigmund Freud*, ed. and trans. James Strachey (London: Hogarth Press, 1950), 18:49. [Hereafter abbreviated as *SE*.]

2. Ibid.

3. Freud, *SE*, 23:148.

4. Freud, *SE*, 21:118–119. On this point, see also the last heading in "Two Encyclopedia Articles," "Recognition of Two Classes of Drives in Mental Life" (Freud, *SE*, 18:258); and *Beyond the Pleasure Principle*: "the efforts of Eros . . . combine organic substances into ever larger unities" (ibid., 42–43).

5. Freud, *SE*, 19:41.

6. Freud, *SE*, 21:122.

7. Freud, *SE*, 19:41.

8. Freud, *SE*, 17:139.

9. Freud, *SE*, 17:115.

10. Freud, *SE*, 7:170.

11. Freud, *SE*, 23:241.

12. Freud, *SE*, 18:251.

13. Ibid., 250. On several occasions, Freud insists on the importance and necessity of a certain psychic plasticity within the psychoanalytic cure. In *An*

Outline of Psychoanalysis, for example, he writes: "A certain psychic inertia, a sluggishness of the libido, which is unwilling to abandon its fixations cannot be welcome to us" (Freud, *SE*, 23:181).

14. Freud, *SE*, 17:116.

15. Freud, *SE*, 23:241–242 (my emphasis).

16. Ibid.

17. Ibid., 243.

18. Ibid., 242.

19. Freud, *SE*, 19:44–45.

20. Freud, *SE*, 18:36.

21. Freud, *SE*, 17:116.

22. Let us recall the definition of inertia: "The principle according to which a heavy body, in the absence of any contrary force, will remain at rest or will move in a straight line."

23. Elasticity names the capacity of certain bodies to regain (at least partially) their original form and volume when a certain force ceases to impinge upon them.

24. In thermodynamics, entropy is a function that defines the degree of disorder within a system that grows as this system evolves toward a new state of relative disorder. It also characterizes the degradation of energy due to an augmentation of this disorder.

25. In certain subjects, Freud writes, "the processes which the treatment sets in motion are so much slower than in other people because, apparently, they cannot make up their minds to detach libidinal cathexes from one object and displace them on to another, although we can discover no special reason for this cathetic loyalty" (*SE*, 23:241).

26. Freud, *SE*, 21:69.

27. Ibid.

28. Ibid.

29. Ibid., 71.

30. Freud, *SE*, 23:229.

31. This formula appears specifically in *The Ego and the Id* throughout Chapter 3, "The Ego and the Superego (Ego-Ideal)" and in "Analysis Terminable and Interminable" (ibid., 224).

32. Ibid., 76.

33. Ibid., 77.

34. Ibid., 78.

35. Ibid., 77–78.

36. Damasio, *Descartes' Error*, 112–113.

37. "Plasticity in the Nervous System," *The Oxford Companion to the Mind*, ed. Richard L. Gregory (Oxford: Oxford University Press, 2004), 623.

38. See Ian Thornton, *Krakatau: The Destruction and Reassembly of an Island Ecosystem* (Cambridge, Mass.: Harvard University Press, 1999). The term "resilience" has a remarkably broad field of application. In information science, resilience designates the quality that allows a system to continue to function correctly even though one or many of its constitutive elements are defective. This quality is called *system resiliency*. More recently, expressions such as "resilient business" or "resilient community" have appeared in American and Canadian publications in order to describe the intrinsic capacity of businesses, organizations, and communities to regain stability—either their initial state or a new equilibrium— allowing them to function in the wake of a disaster or in the presence of continuous stress. According to the same logic, one may speak of "resilient" societies, ethnic groups, languages, or systems of belief.

39. Cyrulnik, *Un merveilleux malheur*, 8.

40. Boris Cyrulnik, transcription of the television program, "La résilience: Origine, définition et principes," www.chez.com/sylviecastaing/resiliens.htm.

41. Boris Cyrulnik, *La naissance de sens* (Paris: Hachette/La Villette, 1991), 41.

42. Cyrulnik, *Un merveilleux malheur*, 186.

43. Ibid., 186.

44. Ibid., 187.

45. Ibid., 21.

46. As opposed to what Freud affirms in *Beyond the Pleasure Principle*: "the patient is, as one might say, fixated to his trauma" (Freud, *SE*, 18:13).

47. Cyrulnik, *Un merveilleux malheur*, 16.

48. Oliver Sacks, *The Man Who Mistook His Wife for a Hat* (New York: Simon & Schuster, 1998), 87.

49. Ibid.

50. Ibid., 88.

51. Oliver Sacks, "Préface" to *L'Homme dont le monde volait en éclats*, trans. Fabienne Mariengof and Nina Rausch de Traubenberg (Paris: Seuil, 1995), 16. The French version of this book also includes *The Mind of a Mnemonist* and the preface includes several pages devoted to the latter text that are not included in the English version of Sacks's preface to *The Man With a Shattered World*. The citations from this text have thus been translated from the French.

52. Ibid., 15–16.

53. A. R. Luria, *Neuropsychological Studies in the USSR*, Part II, Proceedings of the National Academy of Science, USA, col. 70 (4), 1973, 1280ff.

54. Sacks, *The Man Who Mistook His Wife for a Hat*, 89.

55. Oliver Sacks, *An Anthropologist on Mars* (New York: Vintage Books, 1995), xviii.

56. Ibid.

57. Oliver Sacks, *Awakenings* (New York: Harper Perennial, 1973), 234. See also *The Man Who Mistook his Wife for a Hat*, which describes how Ray "has achieved what Nietzsche liked to call 'The Great Health'" (101), finding a way of using his drug, Haldol, that leaves just enough room for "his Tourette's." During the week, he is his "Haldol self," while, during the weekend, he is his "Tourette's self."

58. Ibid., 229.

59. A. R. Luria, *The Making of a Mind*, cited by Oliver Sacks in his preface to Luria, *The Man With a Shattered World: The History of a Brain Wound*, trans. Lynn Solotaroff (Cambridge, Mass.: Harvard University Press, 1987), x.

60. Ibid., xi.

61. Ibid., xii.

62. Ibid., xiv.

63. Sacks, *Awakenings*, 285.

64. Ibid., 228.

10.   TOWARD A PLASTICITY OF THE COMPULSION TO REPEAT

1. Sigmund Freud, *The Standard Edition of the Complete Psychological Works of Sigmund Freud*, ed. and trans. James Strachey (London: Hogarth Press, 1950), 18:53. [Hereafter abbreviated as *SE*.]

2. André Green, *The Work of the Negative*, trans. Andrew Weller (London: Free Association Books, 1999), 83.

3. Freud, *SE*, 18:53–54.

4. Ibid., 54.

5. Freud, *SE*, 19:40.

6. Ibid., 41.

7. See Freud, "The Economic Problem of Masochism" (1924), ibid., 157–170.

8. Ibid., 41–42.

9. Ibid., 42–43.

10. Ibid., 44–45.

11. Freud, *SE*, 18:7.

12. Ibid., 11.

13. Ibid.

14. Ibid., 16.

15. Ibid., 20.

16. Ibid., 29–30.

17. Ibid., 31.

18. Ibid., 33.

19. Ibid., 32.

20. Ibid., 62.

21. Ibid.

22. Ibid.

23. According to the law of the "Nirvana principle" discussed earlier.

24. Ibid., 32.

25. Ibid.

26. J. Laplanche and J.-B. Pontalis, *The Language of Psychoanalysis*, trans. David Nicholson-Smith (New York: Norton, 1974), 172.

27. Boris Cyrulnik, *The Whispering of Ghosts*, trans. Susan Fairfield (New York: Other Press, 2005).

28. It is in her discussion of Paul de Man's notion of a "break within the system" that Caruth proposes that of a "deathlike break" within the framework of trauma. See Cathy Caruth, *Unclaimed Experience: Trauma, Narrative, and History* (Baltimore: Johns Hopkins University Press, 1996), 87.

29. Baruch Spinoza, *Ethics*, in *The Spinoza Reader*, ed. and trans. Edwin Curley (Princeton: Princeton University Press, 1994), 221–222. My emphasis.

II. THE SUBJECT OF THE ACCIDENT

1. Marcel Gauchet, *L'inconscient cérébral* (Paris: Seuil, 1992), 36.

2. Michel Foucault, *The History of Sexuality, Volume 1: An Introduction*, trans. Robert Hurley (New York: Random House, 1978), 11.

3. Ibid., 98.

4. Michel Foucault, *Dits et écrits II, 1976–1988* (Paris: Gallimard, 2001), 105.

5. Foucault, *The History of Sexuality*, 157.

6. Michel Foucault, "Nietzsche, Genealogy, History," in *Language, Counter-Memory, Practice: Selected Essays and Interviews*, ed. Donald F. Bouchard (Ithaca, N.Y.: Cornell University Press, 1980), 150.

7. Michel Foucault, *Aesthetics, Method, and Epistemology*, ed. James G. Faubion (New York: The New Press, 1998), 205.

8. Ibid., 206.

9. Jacques Derrida, *Without Alibi*, ed. and trans. Peggy Kamuf (Stanford: Stanford University Press, 2002), 244.

10. Ibid.

11. Ibid., 244–245.

12. Ibid., 245.

13. Ibid., 246.

14. See my analyses in Chapter 2 devoted to lesions of the frontal lobe, the region essential to the "emotional brain."

15. Derrida, *Without Alibi*, 257.

16. Ibid.
17. Ibid., 258.
18. Ibid., 254.

CONCLUSION

1. Jacques Lacan, *Le Séminaire, tome X: L'angoisse* (Paris: Seuil, 2004), 43.
2. Ibid.
3. Sigmund Freud, *The Standard Edition of the Complete Psychological Works of Sigmund Freud*, ed. and trans. James Strachey (London: Hogarth Press, 1950), 14:286.
4. See Chapter 7.
5. Daniel Widlöcher, *Les nouvelles cartes de la psychanalyse* (Paris: Odile Jacob, 1996), 262, 267.
6. Françoise Davoine and Jean-Max Gaudillière, *History Beyond Trauma: Whereof One Cannot Speak, Thereof One Cannot Stay Silent*, trans. Susan Fairfield (New York: Other Press, 2004), 6.
7. Ibid., 49.
8. As Ruth Leys suggests in *Trauma: A Genealogy* (Chicago: University of Chicago Press, 2000), 37–39.

American Psychiatric Association. *Diagnostic and Statistical Manual of Mental Disorders*, 3rd edition (DSM-III). Washington, D.C.: American Psychiatric Association, 1980; revised 3rd edition (DSM-III-R), 1987; 4th edition (DSM-IV), 1994.

Andreasen, Nicolas. "Defining the Phenotype of Schizophrenia: Cognitive Dysmetria and its Neural Mechanisms." *Biological Psychiatry* 46, no. 7 (1999): 908–920.

Ansermet, François, and Pierre Magistretti. *The Biology of Freedom: Neural Plasticity, Experience, and the Unconscious*. Translated by Susan Fairfield. New York: Other Press, 2007.

Barrois, Claude. *Les névroses traumatiques*. Paris: Dunod, 1988.

———. "Souvenir de l'enfer, enfer du souvenir." In *Stress, psychiatrie et guerre*, 143–149. Paris: Servier, 2002.

Bayley, John. *Elegy for Iris*. New York: Picador, 1999.

Beckett, Samuel. *Happy Days*. New York: Grove Press, 1961.

Bettelheim, Bruno. *The Empty Fortress*. New York: Free Press, 1967.

Blond, Olivier. "Une retombée en enfance?" *La Recherche* 10 (January–March 2003): 80–81.

Caruth, Cathy. *Unclaimed Experience: Trauma, Narrative, and History*. Baltimore: Johns Hopkins University Press, 1996.

Castel, Robert. *Les métamorphoses de la question sociale*. Paris: Gallimard, 1995.

Changeux, Jean-Pierre. *Neuronal Man*. Translated by Laurence Garey. Princeton: Princeton University Press, 1997.

Crocq, Louis. *Les traumatismes psychiques de guerre*. Paris: Odile Jacob, 1999.

Cyrulnik, Boris. *La naissance du sens*. Paris: Hachette/La Villette, 1991.

———. *Un merveilleux malheur*. Paris: Odile Jacob, 2002.

———. *The Whispering of Ghosts*. Translated by Susan Fairfield. New York: Other Press, 2005.

Damasio, Antonio. *Descartes' Error: Emotion, Reason, and the Human Brain*. New York: Penguin, 2005.

―――. *The Feeling of What Happens*. San Diego: Harcourt, 1999.

―――. *Looking for Spinoza: Joy, Sorrow, and the Feeling Brain*. New York: Harcourt, 2003.

Davoine, Françoise, and Jean-Max Gaudillière. *History Beyond Trauma: Whereof One Cannot Speak, Thereof One Cannot Stay Silent*. Translated by Susan Fairfield. New York: Other Press, 2004.

de Certeau, Michel. *Heterologies: Discourse on the Other*. Translated by Brian Massumi. Minneapolis: University of Minnesota Press, 1986.

Deleuze, Gilles. *Difference and Repetition*. Translated by Paul Patton. New York: Columbia University Press, 1994.

―――. "The Exhausted." In *Essays Critical and Clinical*, 152–174. Translated by Daniel Smith and Michael A. Greco. London: Verso Books, 1998.

Derrida, Jacques. "Psychoanalysis Searches the State of Its Soul." In *Without Alibi*, 238–280. Translated by Peggy Kamuf. Stanford: Stanford University Press, 2002.

Eissler, Kurt R. *Freud as an Expert Witness: The Discussion of War Neuroses between Freud and Wagner-Jauregg*. Translated by Christine Trollope. Madison, Conn.: International Universities Press, 1986.

Fenichel, Otto. *The Psychoanalytic Theory of the Neurosis*. New York: Kegan Paul, Trench, Trubner & Co., 1946.

Foucault, Michel. *Dits et écrits I: 1954–1975*. Paris: Gallimard, 2001.

―――. *Dits et écrits II, 1976–1988*. Paris: Gallimard, 2001.

―――. *The History of Sexuality: An Introduction*. Vol. 1. Translated by Robert Hurley. New York: Vintage Books, 1978.

Freud, Sigmund. *A Phylogenetic Fantasy: Overview of the Transference Neuroses*. Edited by Ilse Grubich-Simitis. Translated by Axel Hoffer and Peter T. Hoffer. Cambridge, Mass.: Harvard University Press, 1987.

―――. *The Standard Edition of the Complete Psychological Works of Sigmund Freud*. 23 volumes. Edited and translated by James Strachey. London: Hogarth Press, 1950.

Gauchet, Marcel. *L'inconscient cérébral*. Paris: Seuil, 1992.

Green, André. *La causalité psychique: Entre nature et culture*. Paris: Odile Jacob, 1995.

―――. *The Work of the Negative*. Translated by Andrew Weller. London: Free Association Books, 1999.

Gregory, Richard L., ed. *The Oxford Companion to the Mind*. Oxford: Oxford University Press, 2004.

Herman, Judith Lewis. *Trauma and Recovery: From Domestic Abuse to Political Terror*. New York: Basic Books, 1997.

Jeannerod, Marc. *Le cerveau intime*. Paris: Odile Jacob, 2002.

———. "Neurosciences et psychiatrie. Attirance ou repulsion?" *Les Temps Modernes* 630–631 (May–June 2005): 68–82.

Kant, Immanuel. *Critique of Pure Reason*. Translated and edited by Paul Guyer and Allen W. Wood. Cambridge: Cambridge University Press, 1997.

Kardiner, Abram, and Cora Dubois. *The Individual and His Society*. New York: Columbia University Press, 1939.

Kardiner, Abram, and Hubert Spiegel. *War, Stress, and Neurotic Illness*. New York: P. B. Hoeber, 1947.

Lacan, Jacques. *Écrits: The First Complete Edition in English*. Translated by Bruce Fink in collaboration with Héloïse Fink and Russell Grigg. New York: Norton, 2006.

———. *Le séminaire, tome X: L'angoisse*. Paris: Seuil, 2004.

———. *The Seminar of Jacques Lacan, Book II: The Ego in Freud's Theory and in the Technique of Psychoanalysis*, 1954–55. Translated by Sylvana Tomaselli. New York: Norton, 1991.

———. *The Seminar of Jacques Lacan, Book VII: The Ethics of Psychoanalysis*. Edited by Jacques-Alain Miller. Translated by Dennis Porter. New York: Norton, 1992.

———. *The Seminar of Jacques Lacan, Book XI: The Four Fundamental Concepts of Psychoanalysis*. Translated by Alan Sheridan. New York: Norton, 1977.

Laplanche, Jean, and Jean-Bertrand Pontalis. *The Language of Psychoanalysis*. Translated by Donald Nicholson-Smith. New York: Norton, 1974.

Lebigot, François, and Nicolas Prieto. "Les soins psychiques précoces en cas de catastrophe" (Early Psychic Care in Case of Disaster). http://bit.ly/uSGQz6.

LeDoux, Joseph. *The Synaptic Self*. New York: Penguin Books, 2002.

Leys, Ruth. *Trauma: A Genealogy*. Chicago: University of Chicago Press, 2000.

Luria, Alexander R. *Higher Cortical Functions in Man*. Translated by Basil Haigh. New York: Basic Books, 1980.

———. *The Making of a Mind: A Personal Account of Soviet Psychology*. Translated by Mike Cole. Cambridge, Mass.: Harvard University Press, 1979.

———. *The Man With a Shattered World: The History of a Brain Wound*. Translated by Lynn Solotaroff. Cambridge, Mass.: Harvard University Press, 1987.

———. *The Mind of a Mnemonist: A Little Book About a Vast Memory*. Translated by Jerome Bruner. Cambridge, Mass.: Harvard University Press, 1987.

———. *Traumatic Aphasia: Its Syndromes, Psychology, and Treatment*. Translated by Douglas Bowden. Berlin: Mouton de Gruyter, 1970.

———. *The Working Brain: An Introduction to Neuropsychology*. Translated by Basil Haigh. New York: Basic Books, 1973.

Malabou, Catherine. *The Future of Hegel: Plasticity, Temporality, Dialectic*. Translated by Lisbeth During. New York: Routledge, 2005.

———. "La naissance de la mort: Hegel et Freud en guerre." In *Hommage à Bernard Bourgeois*, 319–331. Edited by François Dagognet and Pierre Osmo. Paris: Vrin, 2000.

———. *What Should We Do with Our Brain?* Translated by Sebastian Rand. New York: Fordham University Press, 2008.

Meyer, Catherine, Mikkel Borch-Jakobsen, Didier Pleux, and Jacques van Rillaer, eds. *Le livre noir de la psychanalyse: Vivre, penser, et aller mieux sans Freud*. Paris: Éditions des Arènes, 2005.

Oppenheim, Hermann. *Die traumatischen Neurosen*. Berlin: V. von August Hirschwald, 1892.

Pommier, Gérard. *Comment les neurosciences démontrent la psychanalyse*. Paris: Flammarion, 2004.

Porge, Erik. "Introduction" to *Freud sur le front des névroses de guerre*, trans. Madeleine Drouin, Anne Porge, Erik Porge, and Anne-Marie Vindras (Paris: Presses Universitaires de France, 1992).

Postel-Vinay, Olivier. "Le cerveau et l'amour." *La Recherche* 380 (November 2004): 32–39.

Prochiantz, Alain. *La construction du cerveau*. Paris: Hachette, 1989.

Proust, Marcel. *Sodom and Gomorrah*. Translated by John Sturrock. New York: Penguin Books, 2002.

Ricoeur, Paul. *Freud and Philosophy*. Translated by Denis Savage. New Haven: Yale University Press, 1970.

Roudinesco, Elisabeth. *Jacques Lacan & Co.: A History of Psychoanalysis in France, 1925–1985*. Translated by Jeffrey Mehlman. Chicago: University of Chicago Press, 1990.

Sacks, Oliver. *An Anthropologist on Mars*. New York: Vintage Books, 1995.

———. *Awakenings*. New York: Harper Perennial, 1973.

———. *The Man Who Mistook His Wife for a Hat*. New York: Simon & Schuster, 1998.

Salmon, Thomas. *The Care and Treatment of Mental Diseases and War Neuroses (Shell Shock) in the British Army*. New York: War Work Committee for Mental Hygiene, 1917.

Schore, Allan N. "Dysregulation of the Right Brain: A Fundamental Mechanism of Traumatic Attachment and the Psychogenesis of Post-Traumatic Stress Disorder." *Australia and New Zealand Journal of Psychiatry* 36 (2002): 9–30.

Simmel, Ernst. *Kriegs-Neurosen und psychisches Trauma: Ihre gegenseitigen Beziehungen dargestellt auf Grund psycho-analytischer, hyponotischer Studien*. Munich: Otto Nemnich, 1918.

Solms, Mark. *The Neuropsychology of Dreams: A Clinico-Anatomical Study*. Mahwah, N.J.: Lawrence Erlbaum Associates, 1997.

———. "Psychoanalyse et neurosciences." *Pour la science* (October 2004): 77–81.

Solms, Mark, and Michael Saling. *A Moment of Transition: Two Neuroscientific Articles by Sigmund Freud*. London: Karnac Books, 1990.

Solms, Mark, and Oliver Turnbull. *The Brain and the Inner World: An Introduction to the Neuroscience of Subjective Experience*. New York: Other Press, 2002.

Spinoza, Baruch. *The Spinoza Reader*. Edited and translated by Edwin Curley. Princeton: Princeton University Press, 1994.

Squire, Larry, and Eric Kandell. *Memory: From Mind to Molecules*. New York: Henry Holt, 1999.

Sulloway, Frank. *Freud, Biologist of the Mind: Beyond the Psychoanalytic Legend*. Cambridge, Mass.: Harvard University Press, 1992.

Thornton, Ian. *Krakatau: The Destruction and Reassembly of an Island Ecosystem*. Cambridge, Mass.: Harvard University Press, 1999.

Thurin, Jean-Michel, Pierre-Henri Castel, and Bernard Perret. "Psychothérapies: Quelle évaluation?" *Esprit* (November 2004): 156–174.

Van der Kolk, Bessel A. "The Body Keeps Score: Memory and the Evolving Psychobiology of Post-Traumatic Stress." *Harvard Review of Psychiatry* 1, no. 5 (January–February 1994): 253–265.

Van der Kolk, Bessel A., Alexander McFarlane, and Lars Weisath. *Traumatic Stress: The Effects of Overwhelming Experience on Mind, Body, and Society*. New York: Guilford Press, 1996.

Vincent, Jean-Didier. *Biologie des passions*. Paris: Odile Jacob, 2002.

Widlöcher, Daniel. "Le parallélisme impossible." In *Communication et représentation: Nouvelles sémiologies en psychopathologie*, 181–206. Edited by Pierre Fedida. Paris: Presses Universitaires de France, 1986.

———. *Les nouvelles cartes de la psychanalyse*. Paris: Odile Jacob, 1996.